THE NATURAL HISTORY OF THE TURTLES OF IOWA

A BUR OAK GUIDE

Holly Carver, series editor

THE NATURAL HISTORY OF THE TURTLES OF IOWA

Terry VanDeWalle and Neil P. Bernstein

Foreword by James L. Christiansen

UNIVERSITY OF IOWA PRESS | IOWA CITY

University of Iowa Press, Iowa City 52242

uipress.uiowa.edu

Printed in the United States of America

Design by April Leidig

Printed on acid-free paper

Library of Congress Cataloging-in-Publication Data
Names: VanDeWalle, Terry, author. | Bernstein, Neil P., 1953– author.
Title: The Natural History of the Turtles of Iowa / Terry VanDeWalle, Neil P. Bernstein.
Identifiers: LCCN 2024008363 (print) | LCCN 2024008364 (ebook) | ISBN 9781609389857 (paperback) | ISBN 9781609389864 (ebook)
Subjects: LCSH: Turtles—Iowa.
Classification: LCC QL666.C5 V33 2024 (print) | LCC QL666.C5 (ebook) | DDC 597.9209777—dc23/eng/20240813
LC record available at https://lccn.loc.gov/2024008363
LC ebook record available at https://lccn.loc.gov/2024008364

Unless otherwise noted, all photos were taken by the authors.

The photo on page iii, an eastern spiny softshell from Bremer County, was taken by Kip Ladage/Ladage Photography.

Jim Christiansen counting ornate box turtle growth rings at Big Sand Mound Nature Preserve in 2006.

We dedicate this book to James L. Christiansen, professor emeritus of biology, Drake University, for his untiring devotion to the science of herpetology. Jim's observations, field studies, research, and herpetological publications—more than seventy of these, thirty-four of which focused on turtles—over nearly four decades have contributed greatly to our understanding of the distribution and ecology of Iowa's reptiles and amphibians.

CONTENTS

FOREWORD

James L. Christiansen

After completing my doctoral studies and marrying my wonderful wife, Sarah, in Albuquerque, I considered where I could make the greatest contribution by working with amphibians and reptiles. That brought me to Iowa, where I grew up and where the factors limiting the survival of reptiles were intense and abundant. An additional attraction was the state's diversity of species intersecting at the northern, western, or eastern edges of their geographic ranges. Also, at the time, no professional herpetologist was studying the state's herpetofauna. Fortunately, Drake University was a good fit, and I spent the next thirty-seven years working there, with a particular emphasis on Iowa's turtles.

A foundation for my studies and those of my students was the collection of more than 2,500 reptile and amphibian specimens assembled by Reeve Bailey—zoology professor at Iowa State University from 1938 to 1943 and later at the University of Michigan—along with his field notes on the many specimens he observed dead on roads that were too badly damaged to preserve. Reeve and I collaborated on summary works on the snakes of Iowa, the lizards and turtles of Iowa, and the salamanders and frogs of Iowa as well as other studies. Our combined work provides a rich source of data over an eighty-four-year period that allows for the analysis of the changes in distribution and the overall trends in survival of Iowa's turtle species now being brought together in this book by Terry VanDeWalle and Neil Bernstein.

Terry joined our research team at Drake University as a graduate student in 1991, where he executed an extremely difficult and time-consuming research project studying factors affecting turtle diversity in riverine habitats throughout Iowa. His study involved setting traps in a few inches of water with seemingly bottomless mud in silted-in river oxbows and searching for trappable habitat in large river impoundments with fluctuating water levels. His research augmented our knowledge of Iowa's reptile and amphibian biodiversity and distribution because he diligently surveyed the frogs, salamanders, lizards, and snakes at his study sites. During his research, he developed a powerful interest in other aspects of turtle biology and a particular interest in eastern massasauga rattlesnakes, an area where he has made a significant contribution in addition to his works on turtle ecology. Much of his knowledge is revealed in his 2022 *The Natural History of the*

Snakes and Lizards of Iowa, published by the University of Iowa Press, required reading for anyone interested in Iowa's wildlife. Terry and I spent many hours in the field together as well as countless hours examining hundreds of Iowa museum specimens, primarily in the Drake University Research Collection.

Neil was an established professor of biology at Mount Mercy University in Cedar Rapids when I met him. His early research, with Antarctic birds, evolved into studies of local birds, prairie communities, fish, and eventually Iowa's ornate box turtles. Our paths crossed at Big Sand Mound Nature Preserve in Muscatine and Louisa Counties, where Drake University had an established study site involving some thirty-two threatened or endangered species in relict sand prairie habitat adjacent to the Mississippi River. Our most adventuresome experience together involved surveys of the deep pools created when the 1993 floods breached the Mississippi River levees. These involved days of extreme heat and traps full of shortnose gars, huge bowfins, and a surprising number of large snapping turtles—all seemingly more interested in inflicting injury than in helping us remove them from the traps. Later we canoed the Mississippi backwaters looking, unsuccessfully, for the source of the unmarked yellow mud turtles that continued to appear at Big Sand Mound Nature Preserve.

Terry and Neil have combined their extensive herpetological and ecological experience with thirty-seven years of my work on Iowa turtles and five years of field notes compiled by Reeve Bailey to produce this beautiful, scientifically accurate, and necessary book on the natural history of Iowa's turtles. It is easy reading and will be appreciated and enjoyed by anyone interested in Iowa's wildlife, especially amphibians and reptiles. The book relates a massive amount of information from peer-reviewed turtle studies and Terry's, Neil's, and my personal experiences; it also documents the changes that have occurred since Euro-American settlement along with some of our adventures while pursuing turtles in Iowa. Because Terry and Neil recognize the importance of verified, accurate locality data, they will provide details of the records used for the distribution maps in the book to qualified researchers upon request.

The herpetofaunal community of Iowa has been under intense pressure as farming methods allowing for larger fields have evolved, pastures have been replaced by row-crop fields, wildlife habitat along waterways and specialized habitats such as sand prairies has been eliminated, marshes have been drained or converted to impounded lakes with predatory fish, rivers and streams have been channelized, and roads have been constructed around nearly every square mile of the state. Road expansion is coupled with increased traffic speeds, greatly decreasing the survival chances of any animal that needs to cross a road to reach feeding, nesting, or overwintering habitat. Terry and Neil's coverage of the challenges and efforts needed to save turtles and their habitats as important components of Iowa's ecosystem compels us to pay particular attention to the conservation section included in the introduction as well as the conservation discussion in every chapter. This book should be owned and studied by every Iowa elected official and by anyone interested in protecting Iowa's turtles.

ACKNOWLEDGMENTS

We thank our friend and colleague Jim Christiansen for his dedication to Iowa herpetology and his many years of research on Iowa's turtles that have contributed greatly to this book. Jim provided his personal field notes along with the field notes of Reeve Bailey—zoology professor at Iowa State University from 1938 to 1943 and later at the University of Michigan—for use in this book, and we thank him for supporting this project and for contributing the foreword.

Very special thanks go to our families. Terry's wife, Wendy, and his daughters, Nicole and Sarah, all at times helped with headstarting, trapping, or radio tracking turtles and have all been a constant source of support during the writing of this book, and Rocky lay beside Terry and eased his stress. Neil's wife, Renate, tolerated turtles in the house and yard along with other stochastic events, and his daughters, Anya and Alexis, helped with field research and turtle care while providing cheerful support.

Many individuals assisted us over the years with fieldwork by trapping turtles, measuring specimens, using radiotelemetry, analyzing data, or providing records. Although we are sure we have forgotten someone—and for that we apologize—we want to thank the following people who helped us in a variety of ways: Donald Becker, Bob Black, Debbie Branch, Sydney Edwards, Rachel Fendrich, Paul Frese, Robb Goldsberry, Scott Gritters, Luke Hodges, Daryl Howell, Drew Howing, Danny Hughes, Jim Jansen, Fred Janzen and his Iowa State students, Karen Kinkead, Dave Kutz, Jeff LeClere, Dale Maxon, Andy McCollum, Mike Olsen, Josh Otten, Elizabeth Owens, Stacey Parks, Kelly Poole, Ryan Rasmussen, Richard "Sandy" Rhodes, Christine Ribic, Becky Richtsmeier, Kevin and Mary Sommerville, Kristi Sutton, Tim Thompson, and Riggs Wilson. In addition, the ornate box turtle chapter benefited from the work of Mount Mercy University and Cornell College students and various volunteers.

Althea Archer produced all the maps and Alexis Bernstein created all the drawings in this book. Both are to be commended for producing such outstanding graphics, which significantly contribute to the quality of the book. Thank you for your hard work.

A huge thank-you to Muir Eaton of the Drake University Research Collection and Kevin Roe of the Iowa State Natural History Collection for providing data and access to turtle specimens, without which the Iowa-specific morphological and reproductive information and much of the distribution data in this book could not have been compiled. We are also indebted to the many researchers who have deposited specimens in the Drake University Research Collection over the years and to Reeve Bailey, whose early work on Iowa herps is invaluable.

We also thank the many other museums and their curators and collection managers who provided us with the Iowa records in their collections, including the Academy of Natural Sciences of Drexel University, American Museum of Natural History, Borror Laboratory of Bioacoustics, Brigham Young University Bean Life Science Museum, California Academy of Sciences, Carnegie Museum of Natural History, Central College, Chicago Academy of Sciences and its Peggy Notebaert Nature Museum, Coe College Collection, Cornell University Museum of Vertebrates, Des Moines Historical Society, Field Museum of Natural History, Florida Museum of Natural History, Illinois Natural History Survey, James Ford Bell Museum, James R. Slater Museum, Los Angeles County Museum of Natural History, Louisiana State University Museum of Natural Science, Luther College, Michigan State University Museum, Milwaukee Public Museum, Mount Mercy University, Museum of Comparative Zoology–Harvard University, Museum of Southwestern Biology at the University of New Mexico, Museum of Vertebrate Zoology at the University of California, Berkeley, National Museum of Natural History–Smithsonian Institution, Northern Illinois University BIOS Herpetology Collection, Royal Ontario Museum, San Diego Natural History Museum, Staatliches Museum für Naturkunde Stuttgart, Texas Cooperative Wildlife Collection, University of Colorado Museum of Natural History, University of Illinois Museum of Natural History, University of Iowa, University of Kansas Biodiversity Institute–Herpetology Collection, University of Michigan Museum of Zoology, University of Nebraska State Museum, University of Northern Iowa, University of Washington Burke Museum Herpetology Collection, University of Wisconsin Zoological Museum, Utah Museum of Natural History, and Western Illinois University.

We also wish to thank the Iowa Department of Natural Resources for giving us access to records from its Natural Areas Inventory Database and the Iowa Multiple Species Inventory and Monitoring Program—and a special thanks to the DNR for its financial assistance with this book and for allowing us to access research areas. We also thank Donald Becker and Christopher Smith at HerpMapper for making Iowa turtle records available. We are especially grateful to Jeff Tamplin for generously providing his personal data for use in the book and to Josh Otten and Danny Hughes for input and discussion on various topics.

Many individuals, organizations, and agencies provided funding over the years for the turtle research that contributed greatly to this book, including Cornell

College, Drake University, Iowa Department of Natural Resources Wildlife Diversity Program, Iowa Department of Transportation, Mount Mercy University, MidAmerican Energy Company, Monsanto Corporation (now Bayer), The Nature Conservancy, R. J. McElroy Trust, Sandy Rhodes, U.S. Army Corps of Engineers, and U.S. Fish and Wildlife Service.

We are indebted to the many people who responded to our requests for photographs, and we appreciate their generosity in allowing us to use them without charge: Mark Anderson, Don Becker, Mark Brown, James Christiansen, Kathryn Deetz, Jeff Faircloth, Dean Hester, Danny Hughes, Kip Ladage, Eric Osmundson, Josh Otten, Elizabeth Owens, Mike Pingleton, Ryan Rasmussen, Chelsea Ewen Rowcliffe, Jeff Tamplin, and Wendy VanDeWalle.

We greatly thank Chad Dolan for making available Iowa Department of Natural Resources annual commercial turtle harvest reports, and Jeff Tamplin and Chad Dolan for reviewing an earlier version of the manuscript.

A sincere thank-you to Holly Carver, a great editor, coach, cheerleader, and friend. Holly is a joy to work with, and her constant and undying support is greatly appreciated.

Finally, we wish to express our sincere thanks to the University of Iowa Press for publishing this book.

THE NATURAL HISTORY OF THE TURTLES OF IOWA

Iowa counties.

INTRODUCTION

The turtle lives 'twixt plated decks
Which practically conceal its sex.
I think it clever of the turtle
In such a fix to be so fertile.
—**Ogden Nash,** "The Turtle"

Turtles play a central role in the histories, cultures, and stories of people worldwide and are arguably the most beloved of the reptiles. In ancient China, the tortoise was a sacred animal—Confucius considered it one of the Four Spiritually Endowed Creatures. In Hindu tradition, the world is supported on the backs of four elephants, themselves resting on the back of a turtle. Sea turtles play an important cultural role for many coastal communities around the world, and numerous Indigenous cultures revere them or consider them ancestors. In Australia, turtles appear as totems and in dreamtime stories and creation narratives of its Aboriginal peoples. In the United States, turtles are a distinguishing feature in the architecture of the U.S. Supreme Court Building where, along with sculptures of lions, rams, and eagles, they represent the slow and deliberate pace of justice.

Many Native American cultures believe turtles represent healing, wisdom, spirituality, health, safety, longevity, protection, and fertility, and turtles figure prominently in some creation stories. North America is referred to as Turtle Island by some Indigenous peoples who believe the land was formed on the back of a turtle. Native Americans had a rich oral tradition in which stories were used to explain what they observed in the natural world. Such is the case with the following story adapted from an Anishinaabe legend of how the turtle got its shell.

Nanaboozhoo (Original Man) had just awakened from a deep sleep that was disturbed by the noisy scolding of blue jays. He was a bit cranky, and besides that he was very hungry. Heading down to the village to find something to eat, he came across some men cooking fish. Their camp was located close to the water, and Nanaboozhoo spied many fish cooking over their fire. The men were happy to give him something to eat but cautioned him that it was hot. Not heeding their warning, he grabbed the fish and burned his hand. He ran to the lake to cool it off in the water but tripped and fell on Mishekae (Turtle), who was sunning on the beach. At that time, Mishekae was not as we know her today. She had no shell and was made of soft skin and bone.

Mishekae complained loudly to Nanaboozhoo to watch where he was going. He felt ashamed of his clumsiness and apologized to her, wondering, "What can I do to make it up to her?" He wanted to do something to help his friend. "I'll have to sit and think it over," he decided, as he returned to his wigwam. Sometime later, he returned to the beach and called for Mishekae, who poked her head through the soft beach mud. Nanaboozhoo picked up two large shells from the shore and placed one on top of the other. He scooped up Mishekae and put her right in the middle, between the shells. Nanaboozhoo took a deep breath and began. "You will never be injured like that again," he said slowly. "Whenever danger threatens, you can pull your legs and head into the shell for protection. You will be able to live in the water as well as on the land, and you will be in your own house at all times." Mishekae approved of her new self and thanked Nanaboozhoo for his wisdom. Moving now in her thick shell, she pushed herself along the shore and disappeared into the water. To this day, she continues to grace Mother Earth, still proudly wearing those two shells.

A turtle's most recognizable trait is indeed its shell. In the 1960s, renowned paleontologist Alfred Sherwood Romer wrote, "Because they are still living, turtles are commonplace objects to us; were they entirely extinct, their shells—the most remarkable defensive armor ever assumed by a tetrapod—would be a cause for wonder." The unique shell is composed of bone that forms as the ribs and vertebrae grow into the outer skin layers, connecting with them to form a solid shell. In turtles, unlike crocodiles, the outer shell is actually integrated into the rest of the skeleton, and the turtle is the only vertebrate whose shoulder and hip bones are located inside its rib cage.

The shell consists of two parts: the carapace (the upper part) and the plastron (the lower part), which are covered by thickened horny or bony plates known as scutes. The carapace and the plastron are connected to each other on the sides either by a bony bridge or by ligaments. Members of the genera *Sternotherus* (musk turtles), *Emydoidea* (Blanding's turtle), and *Terrapene* (box turtles) have a single transverse hinge on the plastron, while the genus *Kinosternon* (mud turtles) has two hinges. The plastral hinges allow the turtle to withdraw into its shell partially or completely, thereby enclosing and protecting its head, limbs, and tail. The shells of softshell turtles have lost their scutes and much of their bony material and are instead covered by tough leathery skin.

Turtles may live on land, in fresh water, in brackish water, or in salt water; their shells and limbs are adapted to the environment in which they live. Those that live partially or entirely on land tend to have higher-domed shells, and the larger species often have elephantine hind limbs to support their weight and shovel-like front limbs for digging. Water-dwelling turtles have flatter and more streamlined shells adapted for moving easily through the water. Those that are semiaquatic have webbing between their toes, whereas the limbs of the fully aquatic sea turtles have been modified into paddle-like flippers.

Turtles are oviparous—their young hatch out of shelled eggs that the female lays in a nest she digs. Romer felt that the reptilian evolution of a terrestrially deposited egg in which the embryo could develop out of the water was one of the key adaptations in vertebrate history, and even fully aquatic species, such as sea turtles, must come onto land to lay their eggs. Unlike amphibians, snakes, birds, and mammals, the sex of the hatchlings of many turtle species is determined not by sex chromosomes at conception but instead after conception by the temperature at which the egg is incubated—known as temperature-dependent sex determination.

Three patterns of temperature-dependent sex determination have been described. In Pattern Ia—common in the semiaquatic pond and box turtles—there is a single transition zone of temperature with low temperatures producing males and high temperatures producing females. Pattern Ib—found in fish and the tuatara—is the opposite, with low temperatures producing females and high temperatures producing males. Pattern II—found in the snapping turtle, eastern musk turtle, and yellow mud turtle—has two transition zones, with females produced at low and high temperatures and males produced at intermediate temperatures.

Turtles grow rapidly when young and more slowly as adults. Individuals of many species grow their entire lives, albeit very slowly in their later years. Turtles are among the longest-lived vertebrates. Aquatic species can live 40 years, some box turtles often live between 40 and 100 years, and giant land tortoises can live 150 years in the wild with some in captivity reaching the age of 200.

Turtles are an ancient group, with the earliest known members appearing approximately 300 to 250 million years before the present during the Late Permian period—before the time of the dinosaurs. Turtles persisted through the Permian mass extinction when over 80 percent of all marine species and 70 percent of all terrestrial species on earth disappeared, lasted through the Triassic mass extinction, and survived the Cretaceous mass extinction approximately 65 million years before the present that ended the 165-million-year dominance of the reptiles. Now these survivors of mass extinctions face the sixth mass extinction the earth has witnessed. What makes the current mass extinction different is its cause. While the other five mass extinctions were the result of natural events, this sixth extinction is due to the actions of a single species—humans—who are altering the earth's life-support systems faster than animals or plants can adapt, particularly vertebrates like turtles that are slow to reproduce.

We both care deeply for turtles and have since we each had small red-eared sliders as pets when we were kids. Though we took different paths toward our fascination, both personally and professionally, with this "curious and improbable creature," as the great herpetologist Archie Carr (1952) called the turtle, we nevertheless ended up in the same place—together writing a book on the turtles of Iowa. And while we remain fascinated by turtles and hope that they can survive for another 250 million years, we are also concerned for their future.

Nanaboozhoo told Mishekae, "Whenever danger threatens, you can pull your legs and head into the shell for protection." Unfortunately, the turtle's shell—that "most remarkable defensive armor ever assumed by a tetrapod," as Romer wrote—offers little protection against the human-caused threats that turtles face today. Whether or not turtles can survive the sixth mass extinction will depend less on their shells and more on us.

TAXONOMY

At present, approximately 360 turtle species are recognized worldwide, ranging in size from the tiny speckled cape tortoise at just 2.5 to 3 inches (6.3–7.6 cm) long to the leatherback sea turtle measuring 6.6 feet (2 m) long and weighing up to 1,500 pounds (680 kg) or more. The largest turtle ever to be documented is *Archelon*, an extinct marine species from the Late Cretaceous period—approximately 100 to 66 million years before the present—measuring 15 feet (4.6 m) long and weighing 2.4 to 3.5 tons (2.2–3.2 metric tons).

Turtles have backbones and are therefore classified as vertebrates. Together with crocodilians, snakes, lizards, and tuataras they belong to the class Reptilia, a group of animals all of which are cold-blooded—ectothermic—and have dry scaly skin, an adaptation to prevent water loss and limit abrasion. While turtles are unable to produce internal metabolic heat, most can maintain a relatively stable body temperature by basking, burrowing, or submerging in water. Turtles are further divided into the order Testudines, which includes freshwater turtles, terrapins, tortoises, and sea turtles; Iowa turtles are placed into the suborder Cryptodira (hidden-necked turtles). Four turtle families occur in Iowa: Chelydridae (snapping turtles), Emydidae (semiaquatic pond and box turtles),

Kinosternidae (mud and musk turtles), and Trionychidae (softshell turtles). In the main body of this book, we have grouped the species and subspecies accounts by family, and each group begins with a brief description of its family.

STATUS AND CONSERVATION

Bounded by the Mississippi and Missouri Rivers, which facilitate the northward and westward dispersal of aquatic turtles, and located at the crossroads of several biological and geological zones, Iowa is at the edge of the geographic range of many of the turtle species in the state, which have larger distributions outside of Iowa. Thirteen turtle species are currently regularly found in Iowa, and turtles are known from all ninety-nine counties. However, the number of species present is not uniform across the state, ranging from two species recorded in some western and southwest Iowa counties to twelve recorded in Muscatine and Louisa Counties in southeast Iowa near the intersection of several major tributaries of the Mississippi River.

The number of species recorded in any given county is likely influenced by the level of search effort in that county—some counties are more heavily searched due to their proximity to colleges and universities, larger cities, natural areas, or known occurrences of certain species. Nevertheless, some distribution patterns do emerge. For instance, turtle species richness—the number of species present—decreases from east to west across the state, with more turtle species found in the eastern half of the state, particularly in counties along the Mississippi, lower Cedar, and lower Iowa Rivers, than in the western half. Terry and Jim Christiansen (1996) found distance westward from the Mississippi River to be a significant predictor of turtle species richness. Land use can also affect species richness. For example, highly agricultural counties tend to have fewer species than those that have more natural areas.

Turtles in Iowa face many threats. In 2000, herpetologist Whit Gibbons and his colleagues identified six significant global threats to reptile populations: habitat loss and degradation, invasive species, environmental pollution, disease, unsustainable use of habitat, and climate change. The Iowa Wildlife Action Plan, developed by the Iowa Department of Natural Resources in 2015, recognizes very similar threats to reptiles in the state. Of the eleven threat categories evaluated in the plan, seven threats—agriculture and aquaculture; transportation and utility corridors; biological resource use, harvesting, and consumption; human intrusion, recreation, and disturbance; natural system modifications; invasive and problematic species; and pollution—are classified as very high for reptiles, and two others—residential, commercial, and recreational development and climate change and severe weather—are classified as high for reptiles. Historically, by far the most serious threat to wildlife in Iowa, including turtles, has been habitat loss and degradation. Since the early 1800s, when the first European settlers arrived in Iowa, native prairies have been reduced by more than 99 percent, prairie pothole wetlands have been reduced by about 95 percent, and, conservatively, over 50 percent of Iowa's original woodlands have been lost.

The largest percentage of this loss is due to the conversion of natural land cover to row-crop production. Early settlers in the state found a land of flat to rolling prairies crisscrossed with forested river valleys. Scattered through the grasslands were thousands of potholes, marshes, and old river oxbows. The rich prairie soil was ideal for crop production, especially in river bottoms; however, the wet soils associated with prairie potholes and other wetlands, along with the frequent flooding and slow drainage of floodwaters in river bottoms, hampered farming efforts and bred mosquitoes, which carried malaria. In order to maximize crop yields, landowners installed drainage ditches and tile lines to "reclaim" marshes and oxbows for agricultural use. In addition, drainage and flood control resulting from the channelization and damming of rivers and streams allowed for the conversion of some poorly drained bottomland to cropland. Beginning around the mid-1960s, farmers switched to more intensive agriculture with larger row-crop fields along with continued wetland drainage, the removal of fencerows, and a much greater use of industrial chemicals. As row-crop production became more intense, wetland drainage and stream channelization also increased. At the same time, urban and suburban areas expanded with associated flood control and navigation improvements. These changes resulted in a loss of habitat that was suitable for turtles, especially for those species with unique habitat requirements.

Stream modification activities, including clearing and log removal, riprapping, widening, deepening, and channel realignment, have been intense across most of Iowa. Likely all of Iowa's streams and rivers have had some level of modification. The more highly modified the waterway, the lower the turtle species richness because the reduction in habitat diversity creates a more homogeneous and simplified environment. Terry and Jim (VanDeWalle and Christiansen 1996) found that northern map turtles, northern false map turtles, smooth softshells, and Blanding's turtles were particularly affected by stream modification because their habitat requirements are so specific. Generalist species, such as snapping turtles, tend to be less affected by stream modification due to their ability to use a wide range of habitats. In southeast Iowa, Neil and Jim studied the impacts of water withdrawals for irrigation and urban development in Muscatine and Louisa Counties, one of the areas of high turtle diversity. These withdrawals, along with the construction of levees along the Mississippi River to prevent flooding, resulted in fewer wetlands and contributed to the decline of rare species such as the yellow mud turtle.

Iowa has over 114,000 miles (183,465 km) of roads, which pose a significant threat to turtles that has only increased over time. When roads fragment habitat or are located between overwintering, nesting, and summer habitats, turtles must make dangerous trips to cross them. Many turtles are killed on roads each year in Iowa, often in late May or June as females make their way to nesting sites. If roads lie between nesting sites and summer habitat, hatchlings must make perilous journeys as they move from their nest sites to marshes or streams. Creating underpasses that allow turtles and other animals to safely cross from one side of a road to the other can help. Terry worked with the Iowa Department of Transportation to design and construct an underpass under Iowa Highway 100 in Linn County for Blanding's turtles, the first such underpass in the state specifically designed for turtles. Monitoring of the turtles in the area after the underpass and the four-lane, high-speed highway were constructed showed that Blanding's, western painted, and snapping turtles all freely used the underpass.

Climate change also has the potential to significantly affect turtle populations. Climate change is occurring faster today than during any period over the past 2 million years. Iowa is already experiencing the effects of climate change in the form of increased frequency of extreme precipitation events, higher temperatures and humidity, increased soil erosion and water runoff, more favorable conditions for the spread of invasive species, and drought, all of which affect turtles to varying degrees. As climate change escalates, long-lived reptiles with low reproductive potential are especially vulnerable because they are unable to rapidly change their behavior or quickly colonize new habitats, if any are available. As the effects of climate change accumulate over time, some turtle species will likely change their current geographic ranges and abundance—some may disappear from the state, while others will expand their ranges into Iowa, especially from states to the south.

One of the more concerning potential effects of climate change on turtle populations is its impact on sex ratios. Because some turtles exhibit temperature-dependent sex determination, even a relatively slight raising or lowering of the average temperature during the egg incubation period may result in a biased sex ratio—all females or all males. Fred Janzen (1994a), studying western painted turtles in Illinois, found that annual offspring sex ratio was highly correlated with mean July air temperature, validating concerns about the effects of climate change on population demography. His work indicated that even a modest increase in average temperature of less than 3.6° F (2° C) may drastically skew sex ratios, and

Sand prairie at Big Sand Mound Nature Preserve in Louisa County.

male offspring in the population of the western painted turtles he studied would effectively be eliminated if the predicted increase in average temperature of 7.2° F (4° C) in central North America within the next hundred years were to occur.

Overharvesting of turtles and their eggs for food is a serious concern in many places in the world, including Iowa, which has a long history of harvesting turtles. This history and the magnitude of the harvest in Iowa are described in the next chapter.

As a result of the many and sometimes dire threats that turtles face in Iowa, as of 2015 the Iowa Department of Natural Resources considered eleven species (85%) to be species of greatest conservation need—species with low and declining populations. Three species (23%) are state-listed as threatened—in danger of becoming endangered—and two species (15%) are state-listed as endangered—in danger of becoming extirpated in Iowa. Two species—the Blanding's turtle and the wood turtle—are under consideration for federal listing under the Endangered Species Act.

On a statewide level, the Iowa Wildlife Action Plan recognizes specific threats to reptiles and the need for reptile conservation and management. The Iowa Multiple Species Inventory and Monitoring Program systematically collects data on reptiles, including their occurrence and distribution across the state, and provides the information to wildlife managers for conservation planning. In 2007, the Iowa Department of Natural Resources dedicated the Southeast Iowa Amphibian and Reptile Conservation Area, the first such area in the nation specifically established for amphibians and reptiles. The 470,000-acre (190,202.5-ha) area covering portions of seven counties in southeast Iowa is home to approximately 70 percent of Iowa's amphibian and reptile species, including a significant percentage of the state's threatened and endangered species. Twelve of Iowa's thirteen turtle species (92%) occur within this area. Private conservation organizations, such as the Bur Oak Land Trust, Iowa Natural Heritage Foundation, and The Nature Conservancy, also manage land in Iowa with an emphasis on imperiled turtles. Such dedicated conservation areas along with preservation and proper management of public natural areas and parks, greenbelts, and wildlife management areas are critical to the conservation of turtles.

Equally as important as the management of public land is the protection and good management of private land. You can take a number of steps to protect and conserve turtles by focusing on the feeding, breeding, and sheltering needs of the species found where you live, including:

1. Learn the natural history of your local turtles. With knowledge come understanding and appreciation. The presence of Iowa's rare turtles in an area is an indicator of good-quality habitat for many other native plants and animals.
2. Watch for turtles on the road and in your yard and avoid running over them with your car and lawn mower. If possible and only if it is safe to do so, move them off the road.
3. Preserve turtle habitat whenever possible by protecting wetlands, ponds, and streams and nearby nesting areas, as well as unique habitats such as sand prairies.
4. Enhance habitat for turtles by planting native plants that will buffer wetlands and nesting areas as well as provide food and shelter for terrestrial turtles.
5. Place or leave dead logs in ponds or wetlands that float in the water to create good basking areas for turtles.
6. Establish and maintain wetlands, oxbows, and ponds to ensure good summer and winter habitat for turtles.

7. Protect and maintain open sandy upland areas adjacent to aquatic habitat as nesting sites for turtles. If possible, exclude mesopredators such as raccoons and skunks from these areas.
8. If you use fire to manage grasslands, burn between November 1 and March 31—outside of turtles' active season—to avoid killing terrestrial turtles, hatchlings, and nesting females.
9. Incorporate personal actions into your daily life to reduce climate change. The Iowa Wildlife Action Plan has recognized that reptiles appear to be one of the groups most vulnerable to climate change.

Conservation of turtles, as with all species, is important. They serve key ecosystem functions as both predators and prey, and they play a part in mineral cycling, seed dispersal, and soil formation, function, and maintenance. Some species serve as keystone species—species that affect the survival and abundance of other species within the communities and ecosystems in which they live. Without keystone species, communities may unravel due to loss of species diversity and ecosystem function. In addition, the iconic turtle is a valuable ambassador in environmental education.

Malcolm McCallum (2021) compared recent declines in turtle biodiversity to past losses of turtles and nondinosaurian reptiles during the most recent mass extinction—the Cretaceous extinction approximately 65 million years before the present—and found that modern extinctions of turtles exceeded the losses at the end of the Cretaceous period. He concluded that if turtle extinctions continue at their current level, massive losses could occur within our lifetime and the entire group could be lost in just a few centuries.

Ralph Waldo Emerson said, "All the thoughts of a turtle are turtles." If turtles are to survive the sixth extinction, our thoughts must be of turtles as well.

ORGANIZATION OF THIS BOOK

We have organized the species accounts by families first, based on phylogenetic order, followed by their respective genera in alphabetical order, then alphabetically by species and subspecies where appropriate. Each account includes the turtle's common name followed by the scientific name and species author—the person or persons credited with the first formal use of the name. When the author's name is enclosed in parentheses, it means the species has been placed in a genus different from the one in the original description. The year in which the species was first described follows the author's name. Organisms are known by both a common name and a binomial—the genus plus the specific epithet—scientific name. Subspecies are known by a trinomial—the genus, the specific epithet, and a subspecific epithet.

Common names have the disadvantage of varying by location and over time. The use of scientific names provides a formal system of naming species and eliminates confusion about the species being discussed. This is not to say that scientific names do not change over time—they do as new information is learned about a species' relationship to other species, and where appropriate we have included information about changes in scientific names in the species accounts. With the exception of the ornate box turtle, we have used Brian Crother et al.'s *Scientific and Standard English Names of Amphibians and Reptiles of North America North of Mexico, with Comments Regarding Confidence in Our Understanding*, 8th edition, 2017, as the standard for common and scientific names.

The scope of this book prevents a complete discussion and citation of all the literature related to each species, although we have tried to find and include the sources most relevant to Iowa turtles. In addition, we have minimized in-text citations to enhance readability. Each species account is organized into twelve main sections as follows.

IOWA STATUS. The conservation status of the species in Iowa.

HISTORICAL SUMMARY. A sampling of the common names for each species in the past, the first specimens collected in the state, and a brief history of the early Iowa literature related to each species. When subspecies are recognized, those whose ranges include Iowa are discussed along with recent changes in taxonomy.

DESCRIPTION. The colors, patterns, and scale arrangement of the species. If juveniles differ from adults, the juvenile coloration is also described. We, along with Jim Christiansen, examined Iowa museum specimens, primarily in the Drake University Research Collection, and gathered data on patterns and coloration of Iowa specimens. We examined only specimens in good enough condition to give us reliable data. We added these data to the morphological data collected by Reeve Bailey and provided to Jim and used the combined dataset in this section.

SIZE. The typical size range for each species, including the straight-line carapace length—from the midpoint of the cervical scute along the midline of the carapace to the posterior-most tip of the carapace—of the largest museum specimen in the Drake University Research Collection. Straight-line plastron length—measured from the anterior edge of the plastron along the midline of the plastron to the posterior edge of the plastron—is also reported for some species. Unless otherwise noted, the record length for each species comes from Robert Powell, Roger

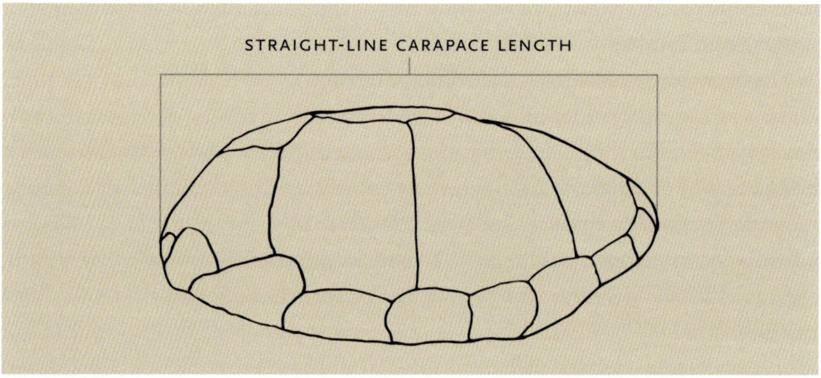

Turtle standard measurement.

Conant, and Joseph Collins's *Peterson Field Guide to Reptiles and Amphibians of Eastern and Central North America*, 4th edition, 2016.

SIMILAR SPECIES. Other Iowa species with a similar appearance along with characteristics that can be used to distinguish separate species.

DISTRIBUTION. A short description of the national range of the species, as shown in the 2016 *Peterson Field Guide to Reptiles and Amphibians of Eastern and Central North America* or other recently published literature, followed by a discussion and a map of the species' range in Iowa. To illustrate changes over time, distribution maps show pre- and post-1960 records that indicate whether a county has only pre-1960 (blue shading) or post-1960 (yellow shading) records or both (green shading). This time break is justified by a hiatus in major herpetological studies in Iowa from about the mid-1950s until 1969, when Jim arrived at Drake University. The source for all records can be tracked to verified specimens in the research and teaching collections of Iowa's colleges, universities, and museums or in museum collections outside Iowa containing Iowa specimens (forty-four total collections), Reeve Bailey's personal field notes, the Iowa Department of Natural Resources Natural Areas Inventory Database and Iowa Multiple Species Inventory and Monitoring Program records, the citizen-science online database HerpMapper, literature citations, or what we consider to be a reliable personal observation verified with a photo. In total, we used 4,898 turtle records from the mid-1800s to 2021 to develop the maps.

Reeve Bailey took extensive notes on turtles he found (mostly road-killed animals) that were not preserved as well as on the many other individuals he

Western painted turtles basking in Bremer County. Photo by Kip Ladage/Ladage Photography.

examined. His field notes on discarded turtles along with pre-1950 literature records are the only records included on the distribution maps that are not represented by a specimen in a collection or supported by a photograph. We are aware that additional observations reported by the public or by qualified observers exist; however, we decided to use only records available to us supported by the above sources for the maps. We are also aware that the distribution maps are a snapshot of current knowledge. Ranges of organisms are changing with climate warming and habitat modifications; therefore, this book provides a basis from which to further expand our knowledge of Iowa turtle biogeography.

HABITAT. Information from literature accounts and Reeve Bailey's field notes and direct field observations by us and Jim.

BEHAVIOR. Temperament, daily or seasonal activities or movements, overwintering habits, and defenses, among other aspects of behavior, from literature accounts, observations in the field by us and Jim, and Reeve Bailey's field notes. In an effort to characterize the seasonality of each species in Iowa, we analyzed the collection/observation dates of the 4,898 Iowa records used for the distribution maps to produce the activity charts in this section. These records may often reflect the schedule of the observer rather than the activity of the species being studied; however, because the records involve many different observers across nearly 150 years at numerous locations across the state, we feel that any search bias is minimal and that the dates provide a good account of the activity period of each species. Because Iowa is at or near the edge of the range of many of the turtle species in the state, the charts show how the seasonality of any given species may differ from that in other parts of its range, especially for widely distributed species.

THREATS. Threats from predators and anthropogenic threats—threats posed by humans.

FOOD AND FEEDING. An examination of the primary foods consumed by each species and its method of acquiring food based on literature accounts, field observations by us and Jim, and Reeve Bailey's field notes. In addition, we examined the stomachs of Iowa museum specimens when available, and a list of any food items we found is provided in this section.

REPRODUCTION. Various aspects of the reproduction of each species, including mating behavior, growth of ovarian follicles (small fluid-filled sacs found on the outside layer of the ovaries that contain immature eggs), ovulation (rupture and release of a matured egg for fertilization when the follicle has grown to the right size), oviposition (egg laying), incubation period, hatching, and size at hatching.

Mating Blanding's turtles in Hamilton County. Photo by Mark L. Anderson, Sanford Museum and Planetarium, Cherokee, Iowa.

Along with information from pertinent literature, Iowa-specific data are based on the internal examination of Iowa female museum specimens by us along with information from Reeve Bailey's field notes.

LONGEVITY. The known lifespan of each species based primarily on Andrew Snider and Kevin Bowler's *Longevity of Reptiles and Amphibians in North American Collections*, 2nd edition, 1992, along with other published accounts.

Also included in the book are a discussion of the exploitation of turtles in Iowa, an illustrated key for identifying adult turtles, a guide to the eggs and hatchlings of Iowa turtles, a glossary of terms, references, a life list, and an index of scientific and common names.

THE EXPLOITATION OF TURTLES IN IOWA

It is time for all men to be told in the plainest terms that there never has existed, anywhere in historic times, a volume of wild life so great that civilized man could not quickly exterminate it by his methods of destruction.
—**William T. Hornaday,** *Our Vanishing Wildlife*

Archaeological evidence shows that our early ancestors began using tortoises and turtles as food at least 2 million years before the present in the Early Pleistocene, and turtles were widely used during the Middle Paleolithic approximately 250,000 to 30,000 years before the present in Europe and the Middle East and during the Middle Stone Age approximately 280,000 to around 50,000 to 25,000 years before the present in South Africa. In North America, Native Americans commonly collected turtles for food, toys, amulets, ceremonial use, and sometimes pets. Turtle shells served as bowls, and other bones were fashioned into utensils.

In Iowa, archaeological records show that snapping turtles and softshell turtles were eaten by some of the state's earliest inhabitants. While teaching at Iowa Lakeside Laboratory in Dickinson County, Neil has noted that among the Native American pottery shards, arrowheads, and tools found by the field archaeology classes, there are always fragments of aquatic turtle shells and bones. In Tama County, members of the Meskwaki Nation, the Sac and Fox Tribe of the Mississippi in Iowa, tell the story of Meshihkêha—the snapping turtle—who was granted the spiritual and physical qualities of a human. He became so egotistical and mischievous and was involved in so many affairs of the heart that the Creator became very angry and changed him back into the snapping turtle we now know, throwing him into the water and saying, "From this day on, your flesh and the flesh of the other members of your clan shall be meat for all men."

As Europeans arrived in North America and spread across the continent, turtles figured prominently in their diet. Early settlements were often near rivers and lakes, and turtles provided advantages over other food sources—settlers could find them, especially nesting females, at predictable times and locations and keep them alive without refrigeration. Unlike larger game species, turtles were relatively easy to catch without great danger or effort. As the human population grew and transitioned to a more urban lifestyle, turtle consumption expanded beyond localized subsistence hunting to the commercial harvest of turtles and their eggs as a salable commodity. As the worldwide demand for turtle meat increased, the commercial turtle-trapping industry in the United States began to export turtles internationally. Commercial and recreational harvesting of turtles for food and the pet trade continues today, and this often unchecked and intense removal has resulted in a global decline of turtles.

By the early 1900s, commercial trade in turtles was common throughout the Mississippi River basin, and historic newspaper accounts from across the state show that it was well underway in Iowa. Numerous articles reported barrels of snapping turtles being shipped from Iowa to restaurants in Philadelphia and Chicago to be sold as "deep sea terrapin" for turtle soup. One wholesale market in Chicago around 1920 was said to have a series of pens in its basement that could hold 2.5 tons (2.3 metric tons) of living snapping turtles, and about 10,000 snapping turtles were handled per year (Clark and Southall 1920). A November 1905 story in the *Decorah Public Opinion* noted that the turtle season had been on for some weeks in the vicinity of Lake City in Calhoun County and that many men had been devoting their entire time to the business. So many turtles had been caught that there was a "famine" of barrels for shipping. The author went on to lament, "Unless some other means can be found to carry the reptiles the harvest will have to end." A 1906 story in the *Marshalltown Evening-Times Republican* reported that over 200 snapping turtles estimated to weigh around 1,200 pounds (544 kg) were caught in Bear Creek in Poweshiek County, and a 1907 story in the same newspaper reported "barrel after barrel filled with big turtles taken from creeks and bayous" near Story City in Story County.

In 1902, the Campbell Soup Company began marketing canned turtle soup, leading to a belief among some Iowans that 300,000 to 400,000 pounds (136–181 metric tons) of Iowa snapping turtles were supplied to the company during the 1920s and 1930s. While this is a colorful story, Scott Gritters and his Iowa Department of Natural Resources colleagues (2013) discovered that the canned soup was really mock turtle soup that contained no real turtle meat at all and was, in fact, made with veal. They went on to conclude that there was no evidence to support the claim that Iowa turtles had been harvested for Campbell's canned soup. Nevertheless, the company did eventually market a frozen turtle soup from 1954 to 1972; however, most of the meat in that product was alligator snapping turtle collected in southern states.

Unrestrained and unregulated turtle harvests continued in Iowa throughout much of the twentieth century. Turtles were taken by hand, turtle hook, turtle trap, hook and line, commercial fishing gear, and other means. Many were taken during the winter by harvesters searching holes in streambanks and other overwintering sites. In a 1942 story in the *Iowa Conservationist* by Everett Speaker, Superintendent of Fisheries for the Iowa Conservation Commission, he wrote, "There are literally tons of snapping turtles in the lakes and streams of Iowa, and this valuable food can and should be harvested." He went on to say, "Harvesting turtles serves a twofold purpose, since it not only furnishes an excellent and valuable food, but also aids in the destruction of a fierce predator." The article also included instructions for how to clean a turtle along with six "tried, tested, and approved" turtle recipes.

In 1987, new rules regulating the commercial harvest of turtles in Iowa were enacted. The rules restricted harvests to the snapping turtle, spiny softshell, smooth softshell, and painted turtle in all waters of the state and allowed for the collection of map turtles only in the Mississippi River and its connected backwaters. A commercial harvester was required to possess a license but was allowed to take, possess, and sell an unlimited quantity of turtles during a continuous open season, meaning that turtles could be taken at any time of the year. No restrictions on the collection of turtle eggs were in place. Recreational collectors holding a valid sport fishing license could take and possess up to 100 pounds (45.4 kg) of live turtles or 50 pounds (22.7 kg) of dressed turtles but could not sell turtles. Importantly, the 1987 rules required commercial turtle harvesters to submit monthly reports to the Iowa Department of Natural Resources, allowing commercial harvesting to be tracked for the first time, though recreational collectors were not required to report their catch. Revisions to the rules in 2006 removed map turtles from the list of species that could be taken and prohibited the collection of turtle eggs from wild nests.

Our review of Iowa Department of Natural Resources annual commercial turtle harvest reports from 1987 through 2021 shows that in 1987, when reporting began, 37 commercial licenses were issued. The number of licenses varied from year to year but generally increased from 1987 through 2012, when 179 licenses were issued. In 1987, 30,807 pounds (13,974 kg) of turtles were reported harvested, and by 2012 that number had risen to over 240,000 pounds (108,862 kg), with snapping turtles making up between 80 and 90 percent (range 62–94%) of the

Snapping turtle in a turtle trap.

catch in most years. Annual harvest by species from 1987 to 2012 ranged from 26,440 to 203,156 pounds (11,993–92,150 kg) of snapping turtles, 338 to 46,469 pounds (153–21,078 kg) of softshells, and 184 to 4,505 pounds (83–2,043 kg) of painted turtles.

When we considered the numbers of turtles by species harvested between 2001 (the first year that harvesters were required to report these numbers) and 2012, the number of individuals harvested per year ranged from 7,676 to 17,209 snapping turtles, 2,217 to 6,953 softshells, and 1,798 to 5,475 painted turtles. Not surprisingly, the lowest numbers for both snapping turtles and painted turtles between 1987 and 2012 occurred during the flood year of 1993, when the long duration of high water made it difficult for harvesters to capture turtles. The greatest change in overall collection numbers occurred between 2006 and 2007, with snapping turtles increasing 65 percent, softshells increasing 91 percent, and painted turtles increasing 112 percent in 2007 compared to 2006. Interestingly, the increase in Iowa collection numbers at this time corresponded to a 79 percent increase (18,457,520 individual turtles) in freshwater turtles exported from the United States in 2007 compared to 2006 (Mali et al. 2014).

The long-term stability of wild turtle populations is affected by a variety of natural and human-caused factors, including disease, habitat destruction, nest depredation, flooding, climate change, and overharvesting. As early as 1992, Iowa DNR biologists noted increases in collection pressure and harvest, leading them to question whether turtles were being overharvested in the state. By 1997, they were recommending establishing a season for the turtle harvest, with a one-month closure during the nesting season and possibly size limits for turtles that could be harvested. As the turtle harvest in Iowa continued to increase throughout the remainder of the 1990s and into the 2000s, the biologists continued to raise concerns and make similar recommendations. Not only were live turtles collected during this time, so too were turtle eggs. The collection of wild turtle eggs in Iowa was unregulated until 2006, and while the number of eggs collected each year is unknown, anecdotal evidence suggests that the harvest may have been massive. In 1992, Terry accompanied Iowa DNR conservation officers on an inspection of a turtle-buying facility in southeast Iowa. In addition to over a hundred live snapping turtles and hundreds of pounds of processed turtle meat ready for sale, the buyer had tray upon tray of turtle eggs, which he claimed totaled approximately 31,000. The eggs were destined to be pickled and sold on the Asian market.

Based on data showing turtle population declines in the United States, in 2009 the Center for Biological Diversity along with two dozen other groups filed emergency petitions to end the unsustainable commercial harvest of freshwater turtles in eight midwestern and southern states, including Iowa. Although the Iowa Natural Resource Commission denied the petition, the Iowa DNR Fisheries Bureau was directed to review the status of turtle populations in the state to assess the need for changes in regulation. Subsequently, a committee of Iowa DNR fisheries and wildlife biologists was tasked with reviewing the turtle program from an unbiased scientific standpoint and recommending regulations and program changes if any were deemed necessary.

In 2013, the Joint Committee on Turtle Harvest issued a report on their findings (Gritters et al. 2013) in which they noted several concerns, including increasing demand for wild-caught turtles for food and pets in addition to fewer wild sources for harvest in the United States. By this time, the commercial turtle harvest in most states surrounding Iowa was prohibited or allowed only under more restrictive regulations, thereby increasing harvest pressure in Iowa. Other concerns included increasing harvest rates over time, loss of habitat, road mortality, nest depredation, and removal of large females from populations. One of the more alarming concerns raised was that the majority of the turtle harvest occurred from May through September, with the peak harvest in June and July during the nesting season. Not only were reproductive females being removed from the population, but in many cases that year's clutch was also lost.

Most Iowans reject hunting other game species, most with much higher reproductive potential, during the nesting and birthing season, understanding that this would lead to population declines over time. Because turtles grow slowly and mature late, and most turtle eggs are depredated within days if not hours of being laid, harvesting turtles during the nesting season is an even graver threat because, although a female may live for decades and build many nests over her lifetime, only a few of her offspring will ever mature into adults. The committee concluded that a year-long unregulated turtle harvest was unsustainable and recommended closing the commercial and sport harvest of turtles during the majority of the egg-laying period.

Due to growing concerns over the turtle harvest, Iowa enacted new rules in 2017. Under those rules, the commercial and recreational collection of spiny softshells, smooth softshells, and painted turtles and the commercial collection of snapping turtles are closed from May 15 to July 15 each year. The recreational collection of snapping turtles remains a continuous open season. The rules also put in place a daily catch limit of four snapping turtles, one spiny softshell or one smooth softshell, and one painted turtle for both commercial and recreational harvesters. Commercial turtle harvesters can possess up to twenty snapping turtles, five spiny softshells or smooth softshells, and five painted turtles. Collecting turtle eggs from wild nests remains prohibited.

According to Iowa DNR annual commercial turtle harvest reports, the overall turtle harvest in Iowa began to decline after 2012, when it reached an all-time high. Nevertheless, in 2016 nearly 110,000 pounds (49,895 kg) were harvested. The 2017 rule change has significantly reduced the harvest. Prior to 2017, the average number of pounds of turtles harvested per year was 142,224 (64,512 kg). Between 2017 and 2021, the annual average dropped to 1,726 pounds (783 kg). However, the total number of pounds harvested has dropped each year since 2016; only 652 pounds (296 kg) were harvested in 2021, all but 16 pounds (7.25 kg) of which were snapping turtles.

Harvesting wild turtles for food is not the only human threat that turtles face; they are also collected, often illegally, to be made into boots, purses, pens, eyeglass frames, bracelets or other jewelry, and traditional medicines among other things. Many turtles are illegally collected for the pet trade. Each year there are stories in the news of hundreds, if not thousands, of illegally obtained turtles being confiscated while in transit to Europe or Asia, many of which die before they arrive. And while it is easy to believe that this occurs only elsewhere in the world, history shows that Iowa is not immune to the black-market wildlife trade. In the late 1990s, Terry assisted Iowa DNR conservation officers who were serving a search warrant in eastern Iowa at the residence of someone suspected of selling reptiles illegally. They found hundreds of living and dead turtles, including 300 eastern box turtles. Most of the live turtles were in poor health due to the conditions in which they were being housed. Among the 300 eastern box turtles transferred to animal shelters for care, only 90 survived. In addition to live turtles, hundreds of shells from many different species were found. While not all of the turtles were native to Iowa, some were, including 10 wood turtles state-listed as endangered.

Humans have a long history with turtles, and many people care deeply about them. Whenever we give turtle talks, a number of people tell us about their lifelong love of and fascination with turtles and describe their pets. We also routinely receive calls from people who have "saved" a turtle that was crossing the road—sometimes bringing it home with them across state lines—or from someone who has purchased turtles that are no longer wanted. While it's understandable and even commendable to enjoy turtles and want to help them, they should never be taken out of the wild and kept as pets.

If you are going to keep a turtle as a pet, be sure that it was legally obtained by purchasing it from a reputable pet dealer or breeder who should be able to provide documentation regarding its origin. Turtles require special care to be kept healthy in captivity, and because they can live a long time—sometimes several decades—keeping them is a long-term commitment. Research and understand the specific needs of the species you plan to keep, including its housing, lighting, food, veterinary care, and other requirements. And should there come a time when you no longer want the turtle, find it a good home; do not release it into the wild even if it is native to the area in which you live. The more we study turtles, the more we realize that the adults possess a wonderful homing instinct. A turtle removed from the wild or "saved" from crossing a road, transported several hundred miles or across the continent, and then released will use most of its energy trying, unsuccessfully, to return to its home. In addition, introducing outside individuals into a population can also introduce parasites, diseases, and nonlocal genes that can harm local populations.

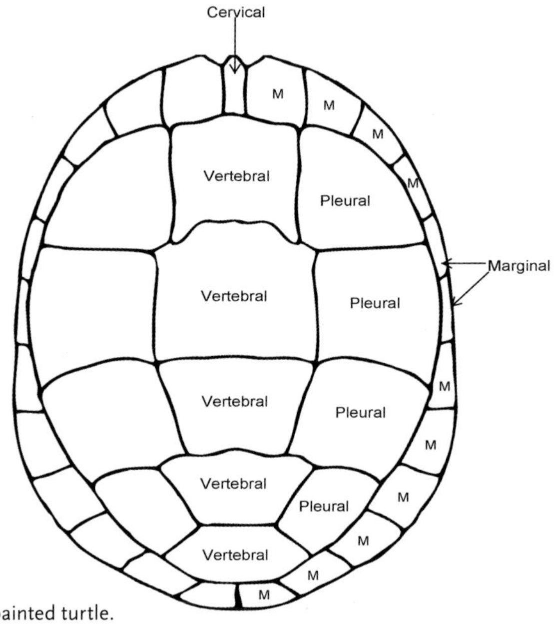

Carapace scutes of a western painted turtle.

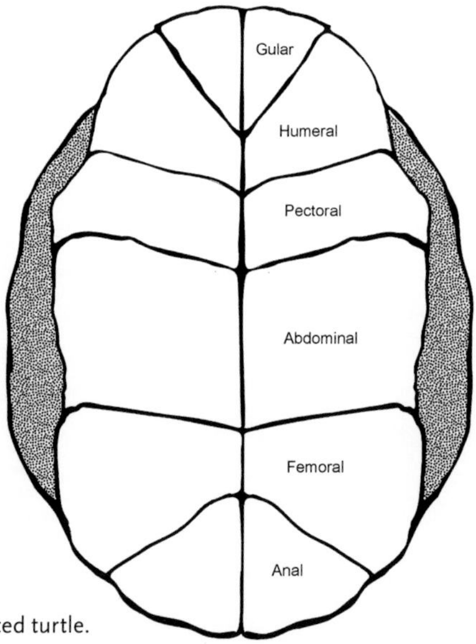

Plastral scutes of a western painted turtle.

IDENTIFYING IOWA'S ADULT TURTLES

The following key is designed to help you identify the turtles in this book. The key is organized into a series of numbered paired choices of characteristics: 1a and 1b, for example. For each pair, beginning with number 1, choose the description that best fits the turtle you want to identify. Move to the next numbered pair indicated at the end of the description. Continue until you reach a description that ends in the name of the turtle. Illustrations of shells and other characteristics are included to aid in identification. See the table that follows this key for a guide to identifying turtle eggs and hatchlings.

1a Carapace flat, soft, covered with leathery skin (not scutes), snout flexible and pointed 2

1b Carapace domed, hard, covered with horny or bony scutes, snout not flexible and not pointed 3

2a Spiny projections on front edge of carapace; horizontal ridge in each nostril; upper side of head, neck, and limbs olive to tan and patterned with dark spots and streaks; throat grayish white with small dark gray markings .... Eastern spiny softshell, *Apalone spinifera spinifera*

2b Front edge of carapace smooth; no horizontal ridge in each nostril; upper side of head, neck, and limbs olive or gray and white or cream below .... Midland smooth softshell, *Apalone mutica mutica*

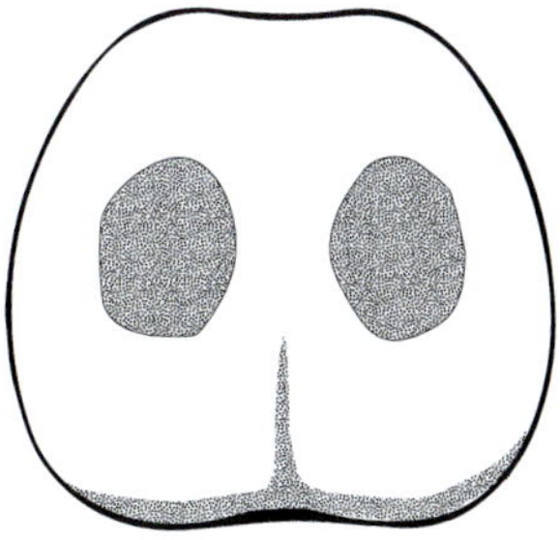

Midland Smooth Softshell

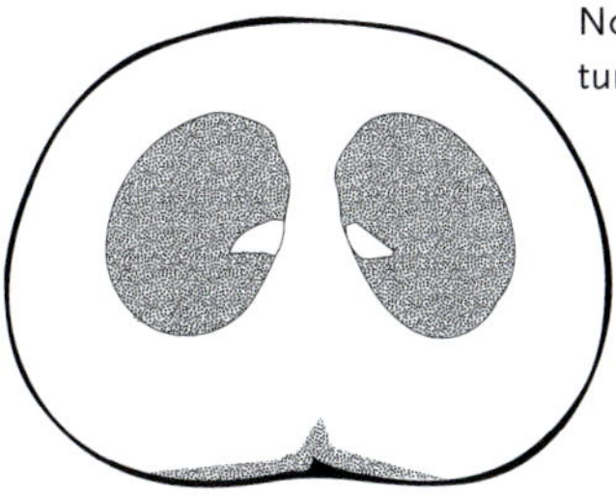

Eastern Spiny Softshell

Nostrils of softshell turtles.

3a Plastron reduced (fewer than 11 scutes), tail long at more than half carapace length with bony projections on top, top of head without large plates, carapace with 3 low keels that may disappear with age .... Snapping turtle, *Chelydra serpentina*

3b Plastron not reduced (11 or more scutes), tail short at less than half carapace length 4

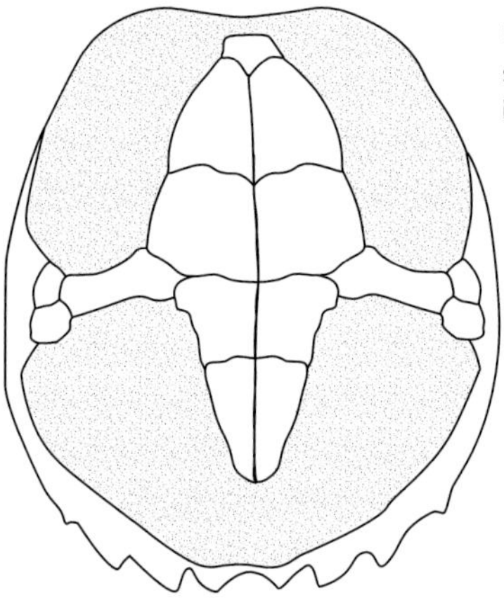

Reduced plastron of a snapping turtle showing 9 plastral scutes. The horizontal bridge is not part of the plastron.

4a Plastron with 11 scutes, pectoral scutes not connected to marginal scutes; carapace smooth, dome-shaped, back edge not serrated 5

4b Plastron with 12 scutes, pectoral scutes connected to marginal scutes; carapace smooth or keeled, back edge serrated or unserrated 6

5a Plastron with 1 poorly developed hinge, white skin showing between scutes, distinctly square pectoral scutes; 2 distinct lines on sides of head; 9th marginal scute same size as 8th marginal scute; tail without prominent spine on tip .... Eastern musk turtle, *Sternotherus odoratus*

5b Plastron with 2 well-developed hinges, no skin showing between scutes, triangular pectoral scutes; 9th and 10th marginal scutes distinctly higher than 8th marginal scute; tail with prominent spine on tip .... Yellow mud turtle, *Kinosternon flavescens*

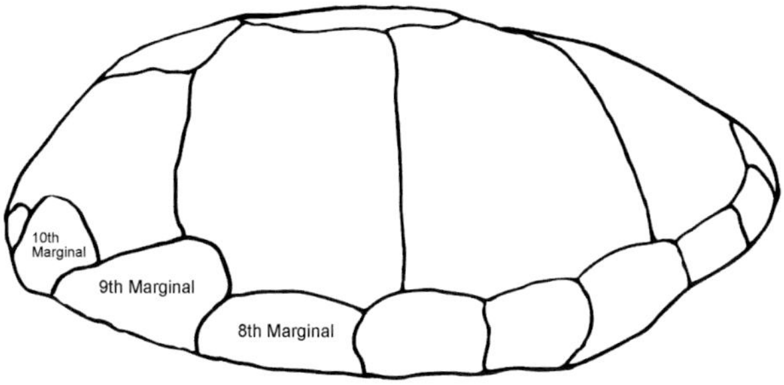

Carapace of a yellow mud turtle showing the distinctly higher ninth and tenth marginal scutes.

6a Plastron hinged: 7

6b Plastron not hinged: 8

7a Chin and throat bright yellow, upper jaw notched; carapace domed, oblong, unkeeled, dark brown to black with numerous tiny yellow or white spots; plastron with 1 hinge......... Blanding's turtle, *Emydoidea blandingii*

7b Chin and throat not bright yellow, upper jaw not notched; carapace domed, unkeeled; radiating yellow lines on carapace and plastron; plastron with 1 hinge; usually 4 toes on each hind foot.. Ornate box turtle, *Terrapene ornata*

8a No light stripes on head, neck, or limbs; carapace with raised pyramid-shaped growth rings on each scute; plastron yellow to cream with dark blotch on each scute....................... Wood turtle, *Glyptemys insculpta*

8b Light stripes on head, neck, and/or limbs.................................. 9

9a Upper jaw with prominent notch; carapace without a keel, back edge not serrated; red bars or crescents on marginal scutes; vertebral stripe absent or poorly developed.. Western painted turtle, *Chrysemys picta bellii*

9b Upper jaw without prominent notch; carapace with or without a keel, back edge serrated... 10

10a Prominent red stripe behind eyes, carapace smooth and weakly keeled, yellow stripe on each pleural scute, plastron with single dark blotch on each scute........................ Red-eared slider, *Trachemys scripta elegans*

10b No prominent red stripe behind eyes; carapace keeled; green to yellow triangular, crescent, or globular mark behind eyes; plastron without large dark blotch on each scute.. 11

11a Green to yellow triangular mark behind eyes not connected to head or neck stripes, low spineless vertebral keel . . . . Northern map turtle, *Graptemys geographica*

11b Yellow mark behind eyes connected to head stripes, vertebral keel topped with low black knobs 12

12a Large yellow square, rectangular, or oval mark behind eyes, large spot under eyes and on lower jaw . . . . Ouachita map turtle, *Graptemys ouachitensis*

12b Narrow yellow backward L–shaped mark behind eyes, no spots on lower jaw . . . . Northern false map turtle, *Graptemys pseudogeographica pseudogeographica*

Northern map turtle | Ouachita map turtle | Northern false map turtle

Head patterns of northern map turtles, Ouachita map turtles, and northern false map turtles.

GUIDE TO EGGS AND HATCHLINGS OF IOWA TURTLES

Carapace	Family	Species	Average Clutch Size	Eggs	Hatchling Characteristics
Hard-shelled	**Chelydridae**	**Snapping turtle**	25 to 45	Spherical, white, 1 to 1.2 inches (2.6–3 cm) diameter, slightly calcified shell	1.1 inches (2.9 cm); carapace rough, light spots on edge; long tail
	Emydidae	**Western painted turtle**	4 to 23	Elliptical, white or cream, 1.2 inches (3 cm) long, leathery shell	1 inch (2.6 cm); nearly round; dorsal keel; large head, eyes, tail; plastron with large dark area
		Blanding's turtle	3 to 22	Ellipsoidal, dull white, 1.5 inches (3.9 cm) long, hard-shelled, nodular surface	1.1 to 1.5 inches (2.9–3.9 cm); carapace dark brown to black, rounded, keeled; plastron black with yellow margins; light yellow chin and neck; long tail
		Wood turtle	3 to 20	Oval, white, average 1.5 inches (3.7 cm) long, leathery thin shell	1.1 to 1.6 inches (2.8–4 cm); carapace rough gray or brown, unmarked dark head and legs, long tail
		Northern map turtle	6 to 20	Ellipsoidal, dull white, 1.3 to 1.4 inches (3.2–3.5 cm) long, soft, easily dented shell	1 to 1.3 inches (2.6–3.3 cm); nearly round; yellow spots behind eyes, map-like pattern on carapace, vertebral keel, dark seams on plastron

Guide to Eggs and Hatchlings (*continued*)

Carapace	Family	Species	Average Clutch Size	Eggs	Hatchling Characteristics
Hard-shelled	**Emydidae**	**Ouachita map turtle**	6 to 15	Elliptical, white, 1.2 to 1.6 inches (3–4 cm) long, leathery shell	1 to 1.4 inches (2.6–3.5 cm); nearly round; rectangular spot behind eyes, pair of spots on jaw, vertebral keel with black knobs, dark pattern on plastron
		Northern false map turtle	8 to 22	Elliptical, white, 1.3 to 1.5 inches (3.2–3.7 cm) long, leathery shell	1 to 1.3 inches (2.6–3.3 cm); nearly round; backward L–shaped mark behind eyes, bright carapace with yellow markings, prominent vertebral keel
		Ornate box turtle	4 to 5	Ellipsoidal, white, 1.4 inches (3.5 cm) long, smooth to finely granular, somewhat brittle shell	1.1 to 1.3 inches (2.8–3.2 cm); carapace light brown with some yellow spotting, tan skin
		Red-eared slider	1 to 30	Oval, white, 1 to 1.6 inches (2.6–4 cm) long, granular shell	1 to 1.6 inches (2.6–4 cm); carapace distinctly green, red stripe behind eyes, dark blotches on plastral scutes

Guide to Eggs and Hatchlings (*continued*)

Carapace	Family	Species	Average Clutch Size	Eggs	Hatchling Characteristics
Hard-shelled	**Kinosternidae**	**Yellow mud turtle**	4 to 6	Elliptical, white, 1 to 1.2 inches (2.6–3.1 cm) long, hard-shelled	0.7 to 1.2 inches (1.7–3.1 cm); black spots at back edge of carapace scutes, yellow on edges of marginals
		Eastern musk turtle	2 to 4	Elliptical, white, 1 inch (2.6 cm) long, hard-shelled	0.7 to 1 inch (1.7–2.6 cm); light stripes on head, carapace black with light spot on each marginal scute, prominent vertebral keel, 2 smaller keels on sides
Soft-shelled	**Trionychidae**	**Midland smooth softshell**	4 to 33	Ovoid to spherical, white, 0.9-inch (2.3 cm) diameter, thick brittle shell	1.2 to 1.5 inches (3–3.7 cm); faint dots and dashes on carapace, underside of carapace brown
		Eastern spiny shell	3 to 42	Spherical, white, average 1 inch (2.6 cm) diameter, thick brittle shell	1.2 to 1.6 inches (3–4 cm); circular spots on carapace, sandpapery feel

FAMILY CHELYDRIDAE

SNAPPING TURTLES

The family Chelydridae is one of the oldest extant turtle families; its fossil record dates back to the Late Cretaceous in North America approximately 100 to 66 million years before the present. Currently, the family is entirely New World in distribution with two living genera: *Chelydra,* the snapping turtles (three species), and *Macrochelys,* the alligator snapping turtles (two species). Iowa has one Chelydridae species, the snapping turtle. While the alligator snapping turtle has been thought by some to occur in the larger rivers of southeast Iowa due to its presence in the Mississippi River just south of Iowa and across the Mississippi River in Illinois, to date no Iowa records have been confirmed.

Members of the family are aggressive with large bodies and heads, powerful jaws with hooked upper beaks, long powerful necks, and long tails crested with large vertical plates. The carapace is well developed, rough, keeled, serrated along the back edge, and connected to a greatly reduced plastron by a thin plastral bridge. Neither the carapace nor the plastron becomes fully bony until late in life, and even then areas near the ends of the ribs may never acquire bone. The limbs are well developed with webbed toes and heavy claws. All members of the family are semiaquatic and are not often seen on land, except females when laying eggs.

small hingeless plastrons, and long saw-toothed tails. The carapace is tan, brown, or black with 3 low keels composed of knobs that may disappear with age; a pattern of light radiating lines may sometimes be present on each scute. The back edge of the carapace is serrated, and 11 marginal scutes are found along each side. A short plastral bridge connects the carapace to a greatly reduced cross-shaped yellowish to tan plastron. The skin may be yellow, tan, gray, black, or reddish, sometimes with small whitish flecks, and the head is large with powerful jaws and well-camouflaged eyes. The neck is long with cream or yellow stripes, small tubercles along the upper surface, and 2 barbels below the chin. The legs are long and strong with webbed toes and thick claws; the long tail—equal to or greater than the straight-line carapace length—is adorned with 3 rows of raised plate-like scales. Hatchlings are darker than adults with a rough carapace, 3 dorsal keels, a mostly black plastron with some scattered white spots, a white spot on the underside of each marginal scute, and a long tail.

Males are larger than females with longer tails and vents that are located beyond the edge of the carapace; females have shorter tails and vents located beneath the carapace.

Plastron of a snapping turtle from Linn County. Photo by Daniel F. Hughes.

SIZE. Iowa's largest turtle, the snapping turtle generally ranges from 8 to 14 inches (20.3–35.6 cm) straight-line carapace length. The largest specimen in the Drake University Research Collection is from Wright County measuring 14.4 inches (36.5 cm) straight-line carapace length, although larger individuals are certainly found in the state, such as a 14.96-inch (38-cm) straight-line carapace length Dickinson County snapper reported by Blanchard (1923) and several very large snappers observed by us over the years. The maximum straight-line carapace length reported for *C. serpentina* is 19.4 inches (49.4 cm), though LeClere (2013) cited a personal communication concerning a 19.75-inch (50.2-cm) straight-line carapace length individual from Emmet County estimated at a minimum of 62 pounds (28.1 kg).

SIMILAR SPECIES. No other Iowa turtle can be confused with the distinctive snapping turtle with its combination of large size, small hingeless plastron, and long saw-toothed tail. Although a few unverified reports of the alligator snapping turtle (*Macrochelys temminckii*) exist from southeast Iowa (VanDeWalle 1993; Christiansen and Bailey 1997; LeClere 2013), no confirmed Iowa museum records, photographs, or literature accounts are known. At present, the alligator snapping turtle is not known to occur in Iowa, but at some point it may enter southeast Iowa as climate warming results in the northward range expansion of many species. Alligator snapping turtles can be distinguished from snapping turtles by a row of scutes located between the marginal and pleural scutes on each side of the carapace.

DISTRIBUTION. Snapping turtles are found from the Atlantic and Gulf Coasts westward to the Rocky Mountains, central Texas, and parts of New Mexico. They have also been introduced west of the Rockies. Along with western painted turtles, snapping turtles have the northernmost distribution among North American turtles, extending into Canada as far north as the Canadian maritime provinces and the southern provinces east of the Rocky Mountains.

In Iowa, snapping turtles are found statewide. In the mid-1940s, the turtle had been documented in 69 counties. In the early 1980s, Jim Christiansen (1981) concluded that construction of farm ponds across the state provided more and ideal habitat for snapping turtles, resulting in an increase in their spatial coverage in Iowa. A review by us of more than 500 post-1960 records indicates that snapping turtles are still frequently found across the state and have been documented in all 99 Iowa counties.

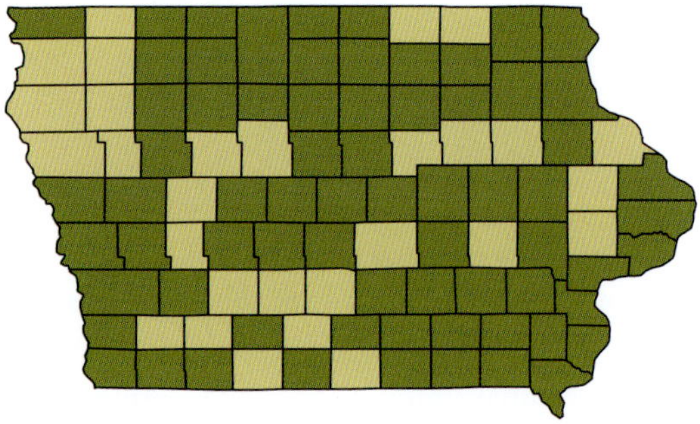

Snapping turtle distribution in Iowa.

Yellow = post-1960
Green = both pre- and post-1960

HABITAT. These turtles are highly aquatic and can be found in nearly any waterbody, including wet ditches, streams, rivers, wetlands, and natural and artificial ponds and lakes. *Chelydra* comes from the Greek *kheludros* meaning "water serpent, a kind of tortoise" in reference to the aquatic habits of the genus. The turtles prefer areas with slow-moving water, soft sand or mud bottoms, and abundant aquatic vegetation and submerged trees and brush. Habitat use changes with age; young snappers prefer shallow water with abundant vegetation while adults will use deeper water. Although shallow water is preferred overall, the turtles can dive to depths of 6.6 to 32.8 feet (2–10 m; Ernst and Lovich 2009).

Reeve Bailey noted habitat for 271 snapping turtles that included woodland pools, woodland streams, woodland lowlands and floodplains, farm ponds, prairie sloughs, drainage ditches, small creeks, dry creek beds, flat uplands, and farm fields with no permanent water. Radio-tracked Iowa snapping turtles were found most often in areas obstructed with brush, logs, and other debris, and individuals were especially associated with beaver lodges (Burken 1978). In Iowa, farm ponds commonly serve as an important source of habitat, especially in areas where no other suitable habitat exists. The Iowa Department of Natural Resources estimates that 110,000 farm ponds are found across the state, and the presence of these ponds has allowed snapping turtles to move into and persist in areas where they might not otherwise have been able to live.

In Polk County, Burken (1978) estimated 7 snapping turtles per acre (17/ha) with an average home range size of 3.3 acres (1.33 ha). Individual turtles had overlapping home ranges, and he felt that home ranges tended to be temporary with the center of the range shifting from one year to the next and, occasionally, within a single year.

BEHAVIOR. The snapping turtle's first response when approached is often to flee or escape into the water. If on land and escape is not possible, the turtle will turn

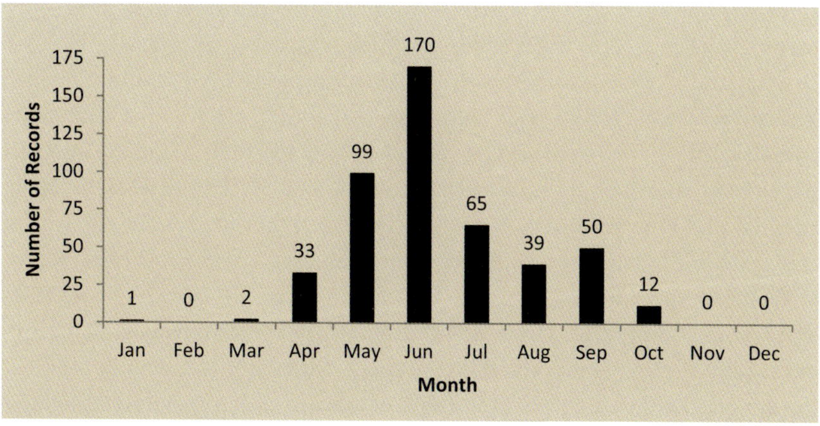

Collection/observation dates of 471 snapping turtles in Iowa.

to face its attacker, lower the front of its carapace, and straighten its rear legs, thus lifting the back of its carapace in an intimidation display. Its throat may swell; the turtle may open its mouth, hiss, and wrap its tail under the carapace; and some may release a pungent odor. Others may stand up on all four legs, lifting the body completely off the ground. The common name "snapping turtle" comes from the turtle's quick temper, aggressiveness, and readiness to strike with surprising speed, sometimes with enough force to lift its body off the ground. The long neck has a surprising reach—*serpentina* is Latin meaning "snake-like" in reference to the long neck—and the powerful jaws are capable of producing a painful and severe bite. Once while trapping turtles in Louisa County, Neil and Jim recovered a trap filled with fish and large agitated snapping turtles. As the turtles were removed from the trap, they tried to bite their captors' legs through the mesh. Although no serious harm was done, they still left bruises.

The collection/observation dates of 471 Iowa snapping turtles show the turtle's primary activity period is from April through October, with the earliest record from Johnson County on January 1 and the latest from Ida County on October 20. Some individuals become active, often under the ice, in March. Jim observed snapping turtles emerging from the water as early as March 9 in Polk County, though most remained in the water until early April, and LeClere (2013) observed a snapping turtle in Boone County on March 23 with its head protruding through the ice.

Snapping turtles are mainly diurnal or crepuscular, spending the day slowly cruising along sand or mud bottoms or basking while floating just beneath the surface of the water with only their eyes and nostrils showing. Less commonly, they may bask out of the water on logs, streambanks, or beaver or muskrat lodges. In Maryland, the turtle is most active in late spring and early summer and much less active from late July through September (Ernst and Lovich 2009). Iowa shows a similar pattern with 57 percent of records in May and June. Males are quite active in May when looking for females. Female activity increases substantially in late May and June during the nesting season, when they are commonly seen on land searching for nesting sites, which likely accounts for the peak in Iowa records in June (36% of records).

As summer progresses and wetlands begin to dry, some individuals may move overland in search of more suitable conditions, while others attempt to wait out the dry period buried deeply in the mud of the wetland or in a terrestrial location. Neil observed dozens of snapping turtles slowly plowing through the mud at Trumbull Lake in Clay County during a prescribed wetland drawdown, presumably searching for deeper water. Burken (1978) radio-tracked snapping turtles in Iowa and found activity during drought was reduced with some turtles stationary while buried in mud or debris for as long as 44 days, and when individuals did move, they tended to move during rain events. Klimstra (1951) observed numerous snapping turtles in southeast Iowa moving overland in August, seemingly in response to a drop in water level in their original habitat. One pair traveled 0.35 mile (0.56 km) from a dried pond to a stream at an average speed of 0.125 mile per hour (0.2 km/hr). Terry frequently sees snapping turtles out his office window during the fall as they move from a shallow constructed wetland to a deeper borrow pond across a mowed lawn in a semi-urban area in Buchanan County. The ability of snapping turtles to move overland promotes the invasion of recently constructed farm ponds or wetlands, often resulting in snapping turtles being the first turtles to colonize.

Snapping turtles begin overwintering by late October, at which time they "mud up" by burrowing into the muddy bottom of a wetland or pond or underneath logs or other submerged debris, or they may take refuge in a beaver or muskrat burrow or lodge. Twenty-six were reported overwintering in a single muskrat burrow in Muscatine County near Muscatine, and twenty were found under a log in a Mississippi River slough (Clark and Southall 1920). They may also wedge themselves under logs or sticks in small flowing streams or in lakes. When fall water temperatures drop below 59° F (15° C), the turtles spend most of their time buried in mud (Ultsch and Lee 1983).

Preferred overwintering sites have some combination of water that is shallow enough for the turtle to reach the surface to breathe without having to swim upward but deep enough that it is unlikely to freeze to the bottom; a location that is likely to be the last to freeze over in the fall or the first to thaw in the spring, cuing spring emergence; mud deep enough for the turtle to bury itself; and some sort of additional cover, such as dead or dormant vegetation, roots, brush, a muskrat lodge, or an overhanging bank (Meeks and Ultsch 1990). Some individuals demonstrate site fidelity to overwintering spots, returning to the same location year after year. Radio-tracked Iowa snappers showed a strong preference for beaver lodges as both summer and overwintering spots (Burken 1978).

THREATS. Raccoons, skunks, mink, and other larger mammals eat snapping turtle eggs and can be significant nest predators. Blanchard (1923) reported a depredated Dickinson County nest found on July 1 that contained 19 eggs, 2 to 3 of which had been eaten. From 30 to 100 percent (mean=70%) of snapping turtle nests were destroyed by predators in Michigan (Congdon et al. 1987), and in South Dakota mammals often visited nests multiple times with complete destruction of 46 to 63 percent of them. Areas of concentrated nests suffered even higher predation rates (Hammer 1969).

Due to their aggressiveness, adults may not be commonly depredated, though river otters, coyotes, and black bears will eat them, as will alligators in the southern United States. Juveniles are easier prey and are eaten by large predatory fish such as bass, wading birds like the great blue heron, bullfrogs, watersnakes, raccoons, and river otters. An endoparasite—a parasite that lives in the host body—was found in Iowa snapping turtles, but the impact on their health, if any, is not known (Wacha and Christiansen 1979). Winterkill is a concern in wetlands where water levels have been reduced due to drought or other causes (Christiansen and Bickham 1989).

Humans pose a significant threat to snapping turtles from drainage and alteration of aquatic habitats, pollution, climate change, and road mortality. Many snappers are killed on Iowa roads each year during the nesting season as females search for places to lay their eggs. As early as the 1930s, Scott (1938) recorded snapping turtle road mortality in Iowa from May through July, with the highest number in June.

Moorman and Carlander (1951) stated that snapping turtles were "unpopular in the pond but welcome at the dinner table!" Until recently, commercial harvesting was likely the biggest threat in Iowa. In the early 1900s, as much as 5 tons (4,536 kg) of snapping turtles were taken during one season in Muscatine County, with 500 to 1,000 pounds (227–454 kg) taken in a single day (Clark and Southall 1920). Based on Iowa Department of Natural Resources records, from 2000 to

2014 commercial turtle trappers harvested from 100,000 to over 200,000 pounds (45,359–over 90,718 kg) of snapping turtles per year in the state. Harvest at this level can have serious population consequences by lowering the reproductive potential of a population through removal of females before they breed (most Iowa females reproduce for the first time at age 7 or 8) and by removal of old females who may produce more than 40 eggs annually. Changes to Iowa's turtle harvest regulations in 2017 made a significant difference. In 2021, commercial turtle trappers harvested fewer than 700 pounds (318 kg) of snapping turtles, illustrating the benefit of good wildlife management. Recreational harvesters are not required to report their catch; therefore, the number of snapping turtles that they take is unknown, though current regulations limit the daily catch to four snapping turtles.

Climate change is an emerging threat that could seriously affect populations in years to come. Because the sex of hatchling snapping turtles is determined by the temperature of the nest, changes in climate that cause nest temperatures to go up or down or that result in a decreased variety of habitats—sun, partial sun, shade—may skew the sex ratio of a population, thereby dooming it in the long term.

FOOD AND FEEDING. The omnivorous snapping turtle will eat any plant or animal that fits into its mouth. Both fresh prey and carrion are readily consumed. Most feeding takes place in the water where the turtle slowly walks or swims along the bottom with its neck outstretched searching for food. Older individuals often use a sit-and-wait strategy—ambushing prey when it gets within range. The turtle quickly thrusts its head and neck forward while rapidly expanding its mouth-pharynx region, creating a low-pressure area that pulls the food into its mouth—pharyngeal gulp. Small prey is swallowed whole, while large prey is held in the jaws and torn into smaller pieces with the front feet before being eaten.

C. serpentina eats various aquatic plants and invertebrates, insects, earthworms, leeches, fish, frogs and toads, small turtles, birds, and small mammals. Slow-moving and sick or injured animals, along with those that have died, are easy prey. Examination of the stomachs of specimens in the Drake University Research Collection found plant material in most stomachs; the fact that some were packed with only plant material like algae and grass suggests that the turtles were actively consuming plants and not just incidentally taking in plants while ingesting something else. Other items included crayfish, a section of a small snake, bird bones, and dragonfly nymphs along with other insect legs, mouth parts, and miscellaneous chitinous structures. Fish scales, vertebrae, and spines were also found along with a complete but partially digested member of the catfish family approximately 4 inches (10.2 cm) long. Nearly all stomachs also had a collection of inedible materials such as mud, sand, sticks, and other debris, likely inadvertently ingested as a result of the sucking action used to capture prey.

Snapping turtles will at times catch small waterfowl, dragging them underwater and drowning them before eating them, leading to the common misconception that snapping turtles depress waterfowl populations. Likewise, they are often accused of lowering game fish populations. However, capture of active, large healthy fish is unlikely, and while predation of young waterfowl does occur, Coulter (1957) estimated that ducklings made up only 10 to 13 percent of the turtles' diet in a season. For comparison, a study of mortality in wood ducks in Mississippi reported that raptors, owls, and herons accounted for greater losses of ducklings than did snapping turtles (Davis et al. 2009).

REPRODUCTION. Jim and Russ Burken (1979) found that Iowa males reach sexual maturity in 4 to 5 years at straight-line carapace lengths of 7.5 to 8.1 inches (19.1–20.7 cm), similar to males elsewhere. For Iowa females, ovarian follicular development begins sometime between their fourth and fifth years at straight-line carapace lengths of 6.2 to 7.5 inches (15.8–19.1 cm), though ovulation does

Snapping turtle nest in Davis County exposed by streambank erosion. Note the flask shape of the nest cavity. Photo by Ryan D. Rasmussen.

not occur until later. While some Iowa females may begin ovulating in their sixth year, most are not sexually mature until 7 to 8 years at straight-line carapace lengths greater than 9 inches (22.9 cm). However, age is not an absolute predictor of sexual maturity; a 6-year-old female had just sexually matured but was larger than many older females, while an 11-year-old female was still sexually immature.

Mating takes place in the water between April and November, most often in late spring and early summer as the water temperature warms. Males may engage in combat at this time for access to females, often with the larger male being the victor. Mating is usually, though not always, preceded by courtship. When a male encounters a female, he may approach her and quickly mount her, or he may crawl or swim around her for a few minutes before mounting. Sometimes the pair may face each other, their snouts almost touching, with each turtle raising the rear of its carapace and lowering the front of its plastron, and they may wave their necks back and forth in opposite directions. Most copulation occurs with the male on top of the female grasping her carapace with all four legs, though plastron to plastron mating may also occur. After mounting, the male begins violent muscular contractions, extends his neck, and may bite the female's head and neck to force her head backward toward her shell. During copulation, the male wraps his tail around the female and elevates her tail while making cloacal contact.

Jim has observed mating in Muscatine County in May and June, and Neil observed it on May 19 and Jim on May 27 in Louisa County. During a Louisa County encounter, loud splashing was heard about 30 feet (9.1 m) from shore and a male was seen grasping a female's carapace with all four feet in what resembled an intense aquatic wrestling match. Only the male's head and sometimes the leading edge of his shell were visible, but occasionally the head of the female would break the surface in front of the male. When the female's head emerged, the male bit at it, driving her underwater. The pair rolled to their sides, plastron to plastron, before separating.

In North America, *C. serpentina* females generally lay eggs between May 15 and June 15 (Ernst and Lovich 2009). Examination of 16 Iowa adult female museum specimens showed that follicular growth begins in late summer or early fall and resumes in the spring with follicles enlarging rapidly. Ovulation in Iowa likely occurs in May when follicles reach approximately 0.8 inch (21 mm) in diameter. The first oviductal eggs were seen in a May 19 female from Louisa County, and by late June oviductal eggs were no longer present, suggesting that egg laying in Iowa likely occurs between late May and mid-June. Consistent with this, during 27 years of field study in Louisa County, Jim and Neil found the earliest snapping turtle nest on May 26, with 9 of 14 observations between June 11 and 16. Obbard and Brooks (1987) noted that environmental temperature was important to stimulate nesting, and Hammer (1969) reported a correlation with rain events.

Most nesting occurs either in the early morning or late afternoon with the entire process taking up to 2 hours. Females dig flask-shaped nests up to 10 inches (25.4 cm) deep with their hind feet and may dig several test holes before selecting a site. Nests are often located in loose sand, loamy soil, or rotting vegetation and, in northern areas like Iowa, they are usually in open sunny locations. Beaver or muskrat lodges may be used at times. Once the eggs have been laid, the female uses her hind feet to fill in the nest. Females may travel up to 1 mile (1.6 km) to find suitable nesting sites, which can put them in danger when they have to cross roadways. In places where suitable nest sites are uncommon, females may be forced to choose sites that are less than desirable, such as the gravel shoulders of roadways, mowed lawns, clay soils, railroad embankments, or under the canopy of a tree. One of Neil's classes from Iowa Lakeside Laboratory in Dickinson County watched a female dig a nest and deposit eggs in hard gravel immediately adjacent to an asphalt driveway at around 8:30 a.m. during the second week of June. To reach the site, she had traveled a minimum of 390 feet (118.8 m) up the steep slope from West Lake Okoboji. Although observed by almost a hundred people over several hours, she was undisturbed and trance-like as the students bent over to watch eggs drop from her cloaca.

Across the snapping turtle's North American range, clutch size varies from 4 to 109 eggs, with most nests containing 25 to 45 eggs (Ernst and Lovich 2009). Counts of shelled oviductal eggs and corpora lutea in Iowa female museum specimens resulted in average clutch sizes of 36 eggs (range 21–52; n = 6) and 31 eggs (range 10–52; n = 8), respectively. Blanchard (1923) reported a nest found in Dickinson County on July 1 that contained 19 eggs. Ninety-three eggs examined in the Drake University Research Collection resembled ping-pong balls—each one nearly spherical, white, and measuring 1 to 1.2 inches (2.6–3 cm) in diameter. The eggshells are less calcified than similarly shaped softshell turtle eggs but considerably more calcified than the eggs of most Iowa turtles (Christiansen and Parmelee 2003).

Eggs hatch from mid-August to October in Iowa after an approximately 75- to 95-day incubation period with hatchlings averaging 1.1 inches (2.9 cm) straight-line carapace length. Well-developed embryos close to hatching and recent hatchlings have been shown to make sounds—principally clicks—within the nest (Geller and Casper 2019a). The precise function of the sounds is not known, but they may be a form of communication between clutch mates that help to synchronize hatching and nest emergence (Spencer and Janzen 2011). Hatchlings emerge from the nest in the fall; they are intolerant of freezing (Storey 2006).

Hatchling snapping turtle from Buchanan County.

After hatching, the young turtles find their way to water through a combination of illumination—light on the horizon—downhill movement, and sometimes by movement toward areas of high humidity (Ernst and Lovich 2009), with larger hatchlings more likely to reach the water sooner. The hatchlings from the Iowa Lakeside Laboratory nest described above emerged in mid-August and immediately headed toward the lake; Christiansen and Gallaway (1984) captured a hatchling moving toward water in Louisa County on September 22; and Terry has commonly found newly emerged hatchlings in late September and early October in Linn and Buchanan Counties.

Like most turtles, snapping turtles lack sex chromosomes, and the sex of hatchlings is determined by the temperature of the nest—temperature-dependent sex determination. The relationship of incubation temperature and hatchling sex is complex, but in general eggs incubated at 68° F (20° C) produce only females, those incubated at 69.8 to 71.6° F (21–22° C) produce both males and females, and those incubated at 73.4 to 75.2° F (23–24° C) produce only males. Eggs incubated at 77 to 82.4° F (25–28° C) produce both sexes but mainly males at the lower end, and eggs incubated at 84.2 to 87.8° F (29–31° C) produce only females.

LONGEVITY. A wild-caught adult survived in captivity an additional 38 years, 8 months, and 27 days, and survival in the wild of 25 to 50 years has been reported.

FAMILY EMYDIDAE

SEMIAQUATIC POND AND BOX TURTLES

The family Emydidae is the largest and most diverse turtle family with two subfamilies, eleven genera, and over fifty species. The family originated in the Eocene epoch approximately 42 to 56 million years before the present, and most species are presently found in North America. However, one genus, *Trachemys,* also occurs in South America, and a second, *Emys,* is exclusive to parts of Europe, western Asia, and northern Africa. Both subfamilies occur in Iowa—Emydinae, which includes the genera *Emydoidea* (Blanding's turtle), *Glyptemys* (wood turtle), and *Terrapene* (ornate box turtle), and Deirochelyinae, which includes the genera *Chrysemys* (western painted turtle), *Graptemys* (map turtles), and *Trachemys* (red-eared slider).

Members of the family usually have a low-arched carapace, though some have a high-domed carapace (e.g., Blanding's turtle and ornate box turtle). The carapace may or may not have a vertebral keel and is usually connected to the plastron by a broad bridge. Turtles in the genera *Emydoidea, Emys,* and *Terrapene* have a movable hinge on the plastron that allows for a closable anterior plastron and varying levels of closure of the posterior plastron. The hinge provides additional, although not foolproof, protection against predators by allowing the turtle to withdraw its head and limbs into the closed or partially closed shell.

The limbs of most emydids are developed for swimming and have, at a minimum, the remnants of toe webbing. The exceptions are the box turtles, whose toes are not webbed, and unlike tortoises they move in a swimming motion across the ground that creates friction between the plastron and the ground, resulting in the plastrons of older individuals being worn smooth. Most emydid species exhibit sexual dimorphism with males of all species having longer and thicker tails and males of some species having concave plastrons. Females of many species are larger than males. All Iowa emydids are semiaquatic with the exception of the ornate box turtle, which is adapted to terrestrial habitats. Due to their frequently colorful appearance, emydids are very popular in the pet trade.

Four subspecies of the painted turtle, *Chrysemys picta,* are currently recognized, only one of which, the western painted turtle, *C. p. bellii,* has a range that includes Iowa.

DESCRIPTION. The western painted turtle is a brightly colored, seemingly hand-painted, turtle whose appearance gives rise to both its common and scientific names. *Chrysemys* is derived from the Greek words *khrysos* meaning "golden," "yellow," or "light green"—probably in reference to the head stripes—and *emys* meaning "turtle"; *picta* is Latin meaning "painted" or "embroidered"—in reference to the delicate shell pattern; *bellii* honors English naturalist and dental surgeon Thomas Bell.

The carapace is olive to black, smooth, oval, flattened, and keelless with no serration along the back edge. The scutes are bordered in yellow or red along the seams, and the marginal scutes have 3 or 4 yellow or red lines that wrap underneath and broaden when viewed from below. A vertebral stripe may be absent or poorly developed. The smooth hingeless plastron is yellow-orange, bright orange, or red and patterned with variably sized black to reddish brown markings, and the plastral bridge is unmarked. The skin is black to olive, and the neck, legs, and tail are striped with yellow or red lines. A yellow stripe extends from the eyes downward and backward to meet a similarly sized stripe from the lower jaw, and the upper jaw has a prominent notch. The pigments in the stripes reflect ultraviolet wavelengths and may aid communication among painted turtles and with other turtle species (Steffen et al. 2015). Hatchlings have a brighter carapace and pattern than adults, a vertebral keel, and a deep crease across the abdominal scute.

Plastron of a western painted turtle from Bremer County. Photo by Kip Ladage/ Ladage Photography.

Western painted turtle from Linn County. Photo by Daniel F. Hughes.

Some males and occasionally some females darken as they age. Two forms of melanism have been described—reticulate, in which a fine network of dark marks occurs on the carapace, and nonreticulate, a general darkening. Reticulate melanism tends to be more common in northern populations and is often associated with mature males, appearing after development of secondary sex characteristics such as elongated front claws (Schueler 1983). The darkening may serve to increase camouflage and, thereby, predator avoidance, or it may aid in thermoregulation—darker individuals warm up quicker (Schueler 1983). Examination of Iowa western painted turtle specimens in the Drake University Research Collection found no instances of reticulate melanism among 77 specimens from eastern Iowa and only 5 instances (all males) among 90 central and western Iowa specimens.

Males are smaller than females, have longer and thicker tails with the vent located beyond the edge of the carapace, and elongated foreclaws; females have shorter tails with the vent located beneath the carapace.

SIZE. Western painted turtles generally range from 4.5 to 8 inches (11.5–20.3 cm) straight-line carapace length. The largest specimen in the Drake University Research Collection is from Madison County measuring 7.9 inches (20.1 cm) straight-line carapace length. The maximum straight-line carapace length reported for the species is 10 inches (25.4 cm).

SIMILAR SPECIES. Among Iowa's other turtles with yellow stripes on the neck and head, the three map turtle species have yellow marks on the head, a distinct vertebral keel, and posterior marginal scutes that are strongly serrated, and they lack red on the plastron. Red-eared sliders have a bright red stripe behind the eyes, posterior marginal scutes that are weakly serrated, and a yellow plastron with dark spots. With the exception of the map turtles and the red-eared slider, all other Iowa turtle species lack yellow stripes on the neck, head, and legs.

DISTRIBUTION. *C. picta* is the most widespread North American turtle and the only one that naturally occurs from the Pacific Ocean to the Atlantic Ocean. Western painted turtles are found in southern Canada from British Columbia to western Ontario; in the United States, they are found in northern Oregon, Washington, and northern Idaho and from Montana, Wyoming, and Colorado east to Michigan's Upper Peninsula, Wisconsin, Illinois, and Missouri. Scattered populations are found in Utah, Arizona, New Mexico, Texas, and portions of Mexico.

In Iowa, western painted turtles are found statewide. In the mid-1940s, the turtle had been documented in 67 counties; at present, records exist for all 99 Iowa counties. In the early 1980s, Jim Christiansen (1981) concluded that construction of farm ponds across the state had increased ideal habitat for western painted turtles and that the turtle was found in large numbers in almost every square mile of Iowa. A review by us of approximately 1,000 post-1960 records indicates that western painted turtles are still frequently found across the state.

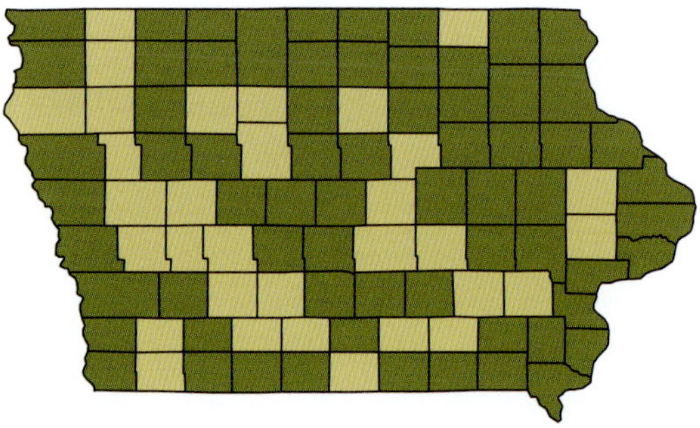

Western painted turtle distribution in Iowa.

Yellow = post-1960
Green = both pre- and post-1960

HABITAT. Western painted turtles prefer slow-moving shallow water with abundant aquatic vegetation, plentiful basking sites, and a soft bottom; they typically avoid water with a fast current. In Iowa, they can be found in nearly any waterbody in the state containing suitable habitat, including quiet portions of rivers and streams, lakes, farm ponds, marshes, sloughs and backwaters, drainage ditches, urban stormwater detention basins, and backyard ponds. Among 131 Iowa western painted turtles observed by Jim, 36 were in marshes, 35 were in quiet pools associated with rivers, 19 were in or near rivers or creeks, 18 were in farm ponds, 18 were in lakes, and 5 were in drainage ditches.

Western painted turtles, along with snapping turtles, are the first turtles to move into newly restored or constructed wetlands in Iowa, sometimes just a few weeks after completion of construction. As a group, painted turtles tend to be more common in landscapes that have plentiful and highly connected waterbodies;

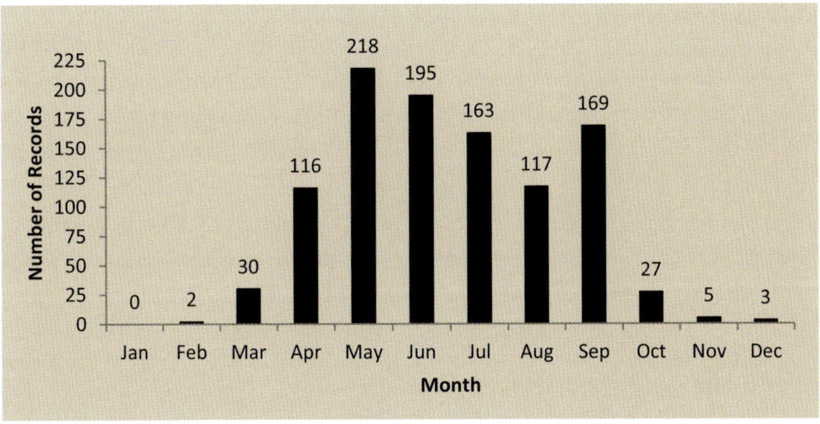

Collection/observation dates of 1,045 western painted turtles in Iowa.

occupied wetlands tend to have little canopy cover, high water levels, large areas, and high connectivity (Cosentino et al. 2010). In the St. Croix River in Minnesota and Wisconsin, western painted turtle abundance was associated with muck bottoms and the number of basking sites, such as snags and rocks (DonnerWright et al. 1999). Habitat use changes with age; hatchlings and juveniles use shallow water areas and move to deeper water as they grow and mature.

BEHAVIOR. Western painted turtles are generally skittish and, if approached while basking, will quickly slip or dive into the water. Interestingly, however, individuals found in areas with less human disturbance appear to be less wary (that is, more fearless) than those found in areas with more human disturbance. Moll (1974) surmised that excessive nervousness—continuously leaving basking sites and swimming away—would waste energy and that calmer individuals may be more successful at feeding and breeding than more nervous ones in areas where predation is not a major concern. If handled, these turtles sometimes bite and hiss, but more commonly they pull into their shells, empty their bladders, and kick and scratch as they attempt to "swim" away.

The collection/observation dates of 1,045 Iowa western painted turtles show the species has been found from February to December in Iowa, though most activity is from late March through September, with a few individuals still seen basking through October and into early November in some years. The earliest Iowa record is one observed basking in a wetland in Linn County that still had a thin coating of ice on the surface on February 19. The turtles may at times be seen swimming under the ice in late winter or early spring, such as one observed by Jim on February 24. They will often start basking on sunny days when the ice is going out, but they do not start feeding in Iowa until sometime in April, likely at water temperatures around 59 to 64.4° F (15–18° C; Ernst and Lovich 2009). By May, they are actively feeding and continue foraging into September, at which time feeding begins to slow down as they prepare for winter. In southwest Minnesota, winter dormancy began in mid-September when the water temperature fell below 68° F (20° C; Ernst and Ernst 1973). The activity season in northwest Iowa may be similar to that in southwest Minnesota, but the season extends later the further south and east you go in the state. The latest Iowa records are from December 18: one found in Wright County in 2001 and four juveniles found in Linn County in 2011.

Sexton (1959) divided annual activity of the closely related midland painted turtle (*C. p. marginata*) in Michigan into five phases: the prevernal stage, beginning with the melting of ice or when the turtles emerge from overwintering; the vernal stage, coinciding with the appearance of aquatic vegetation on the surface of the water and the time of mating; the aestival stage, beginning with nesting, extending through summer, and ending when the turtles begin to return to their overwintering areas; the autumnal stage, occurring throughout the fall until ice forms; and the hiemal stage, when the turtles overwinter under the ice.

Western painted turtles are almost entirely diurnal and, when not foraging, they are basking. While solitary individuals bask, it is common to see large groups basking together, sometimes on top of other turtles, including snapping turtles. They can often be seen basking on floating mats of vegetation; muskrat lodges; partially submerged logs, rocks, or stumps; sandbars or the banks of streams, ponds, or lakes; or even any artificial structures that may be in the water. Daily activity begins around sunrise with the turtles basking for several hours before starting to search for food. After feeding for a while, they once again bask before feeding again in the late afternoon and early evening. Females may bask longer than males at certain times of the year—for example, during the nesting season—leading some to surmise that they bask longer and more frequently prior to nesting to support egg development. However, two studies of painted turtle basking behavior in Ontario do not support this assertion (Lefevre and Brooks 1995; Krawchuk and Brooks 1998). In contrast, in another Ontario study, females basked more than males in mid-August, with the authors attributing the differences between the sexes to the females' need for increased thermal energy uptake during egg development (Carrière et al. 2008).

In terms of thermoregulation, painted turtles select areas that are on average warmer than surrounding areas, behaviorally avoid extreme temperatures, and spend more time basking early in the season than in the rest of the active season (Edwards and Blouin-Demers 2007). Basking can lead to overheating and evaporative water loss through the skin and lungs. Up to two-thirds of a reptile's water loss can be through the skin, depending on the habitat (Bentley and Schmidt-Nielsen 1966). Ernst (1972) validated the relationship between water loss and habitat for painted turtles by showing that individuals lost 8.2 to 11 percent of their body weight through evaporative water loss when exposed to summer temperatures of 50 to 84° F (10–29° C) and relative humidities of 45 to 95 percent.

Western painted turtles can be quite mobile. Most movement is associated with females searching for nesting sites. When their ponds dry in southeast Iowa, these turtles, unlike some turtle species, tend to move to other ponds rather than burying themselves in the mud or nearby sand dunes, moving back to their original ponds when the water returns (Christiansen and Bickham 1989). Home range size has not been reported for Iowa, but in Michigan average home range sizes during the active season were 3.7 acres (1.5 ha) for juveniles, 4.4 acres (1.8 ha) for females, and 7.2 acres (2.9 ha) for males (Rowe and Dalgarn 2010). Individual turtles tended to inhabit the water's edge but frequently used open water habitats, moving up to 164 feet (50 m) within a four-hour period. Several studies indicate that painted turtles use smaller core areas within their daily home range. Abundance of habitat and food results in the use of larger areas; however, home ranges and distances moved vary between locations, habitats, and local climate. At night, the turtles usually sleep underwater.

These turtles generally overwinter in the soft bottoms of ponds, lakes, and marshes, in muskrat lodges, or beneath undercut streambanks, usually in water less than 6.6 feet (2 m) deep but deep enough to prevent ice forming to the bottom. While they are submerged during the winter, oxygen is extracted from the water and carbon dioxide from the blood through a combination of cloacal and buccopharyngeal respiration. In Ontario, painted turtles were active under the ice as long as the water was oxygenated; however, once anoxic—containing no oxygen—conditions arose, activity ceased (Taylor and Nol 1989).

Well-oxygenated water is important, but anaerobic—existing in the absence of oxygen—metabolism also occurs. Western painted turtles are well adapted for anaerobic metabolism and incorporate a collection of cellular, physiological, and behavioral responses to balance and prioritize energy needs, manage metabolic wastes, and maintain vital body processes (Alderman et al. 2021). As the turtles' metabolic and heart rates drop, glycogen—stored carbohydrate—is used to support organ and tissue systems, and the blood becomes acidic due to an accumulation of lactic acid. To counter the buildup of lactic acid, calcium carbonate is mobilized from skeletal bones and the shell into the blood to buffer the acid, and lactic acid is simultaneously stored in those bony tissues. Because of their smaller size, hatchlings are less able to buffer lactic acid, but they often remain in the nest over the winter where anoxic conditions are not a threat. However, hatchlings overwintering in the nest are susceptible to freezing, which is prevented by an accumulation of antioxidants and ice-nucleating agents—cryoprotectants—in the tissues, thereby providing the ability to survive the formation of ice in the body.

THREATS. Western painted turtles fall prey to a large number of predators. Eggs and hatchlings are eaten by garter snakes, ground squirrels, skunks, badgers, and foxes. Juveniles are consumed by giant water bugs, bullfrogs, snapping turtles, northern watersnakes, great blue herons and other wading birds, hawks, screech owls, American crows, mink, opossums, and river otters. The greatest natural predator, however, is likely the raccoon, which preys on large numbers of eggs, hatchlings, juveniles, and adults. In many places, nest predation can be nearly 100 percent. Timing the laying of eggs to coincide with rainfall may reduce the chance of predation (Bowen and Janzen 2005).

While studying the food habitats of muskrats in Iowa, the great wildlife biologist Paul Errington (1941) noted the "flesh-eating tendencies" of muskrats in lakes in northwest Iowa where they consumed western painted turtles. He observed, "Turtles seem most likely to serve as a fall and winter source of food about Iowa lakes when large numbers [of turtles] are torpid or sick or floating dead along the shores, either in open water or under the ice. Ordinarily, the appendages are eaten away from the shell and the latter left nibbled around the edges; but pieces may be gnawed out of a two-to-five-inch carapace or plastron to facilitate removal of the edible material within."

Humans pose threats to western painted turtles through habitat loss and water pollution, which reduce the quantity and quality of the available habitat. Road mortality can be a significant threat where roads bisect or are adjacent to wetlands or where roads are located between wetlands and nesting areas. The threat of roads to western painted turtles in Iowa was demonstrated as early as the 1930s, primarily during the nesting season (Stoner 1936; Scott 1938). Large numbers of these turtles are still hit on roads each year in Iowa. Winterkill is a concern in waterbodies where water levels have been reduced due to drought or where there is an excessive growth of aquatic plants, often the result of nutrient pollution. In such a situation, ice forms over shallow water and then snow accumulates on top of the ice, blocking the sunlight. As the aquatic plants die and decompose, the dissolved oxygen level in the water is depleted, resulting in the death of turtles and other aquatic organisms. During times of low water, turtles buried in the mud may freeze or die from desiccation. Christiansen and Bickham (1989) documented such winterkill of western painted turtles in Louisa County, which has also been witnessed by Neil in the same locations.

Commercial and recreational harvesting of western painted turtles is allowed in Iowa, and in the past it may have posed a threat to some populations in the state. Based on Iowa Department of Natural Resources records, from 1987 to 2016 commercial turtle trappers harvested from 184 to over 4,500 pounds (83–over 2,043 kg) of western painted turtles per year in the state (mean = 2,136 pounds [969 kg]). Changes to Iowa's turtle harvest regulations in 2017 made a significant difference. From 2017 through 2021, only 16 pounds (7 kg) of western painted turtles were reported by commercial trappers. Recreational harvesters are not required to report their catch; therefore, the number of painted turtles taken by recreational harvesters is unknown, though under current regulations recreational and commercial collection of painted turtles is closed from May 15 to July 15; outside of these dates, the rules limit daily catch to one painted turtle for both recreational and commercial harvesters. The collection of wild turtle eggs in Iowa was allowed until 2006; this undoubtedly included western painted turtle eggs, though because harvesters were not required to report collection of eggs, the exact toll is unknown.

FOOD AND FEEDING. *C. p. bellii* is an omnivorous aquatic forager that consumes a wide variety of plants and animals. Known foods, among many others, include algae, bryophytes (mosses and liverworts), vascular plants and seeds, earthworms, slugs and snails, crayfish, freshwater sponges, and insects—aquatic larvae are most often consumed. Vertebrates including fish, amphibians, and birds are also eaten, probably as carrion. Larger prey is held in the jaws and shredded with the front feet. Moving prey is detected faster than stationary prey, and the turtles may also forage along the bottom, using their heads and forelimbs to disturb the vegetation to flush prey for capture. They may also employ a sit-and-wait strategy, ambushing passing prey. Individuals have also been observed skimming the surface of the water, trapping fine food items by allowing water to flow into their throats—a process known as neustophagia (Belkin and Gans 1968). Anthony Quinn and Jim (1972) found that Iowa western painted turtles grew faster in habitats with softer organic bottoms—more nutritive detritus—than in those with a sand or gravel bottom. The closely related eastern painted turtle (*C. p. picta*) has been shown to be effective at dispersing seeds of aquatic plants in their feces (Padgett et al. 2010); the same is likely true for the western painted turtle.

REPRODUCTION. Males reach sexual maturity at around 2 to 4 years at 2.8 to 3.7 inches (7–9.5 cm) straight-line plastron length, though maturation in northern populations could be as late as the sixth year. Females mature later at around 6 to 10 years at 3.8 to 5 inches (9.7–12.8 cm) straight-line plastron length. Reproductive maturity is correlated more with size than with age; both size and age of maturity increase with latitude and elevation, and individuals grow larger in areas with higher mean annual temperatures and more frost-free days (Iverson and Smith 1993).

Male and female reproductive cycles in Wisconsin were described by Christiansen and Moll (1973) and are likely similar in Iowa. They considered three overlapping stages in the female cycle: phase 1, August through June, a period of follicular enlargement during which the cycle pauses while the female overwinters and resumes and accelerates the following spring; phase 2, late May through the first three weeks of July, a period during which eggs are ovulated and laid; and phase 3, the second week of July through August, characterized by ovarian dormancy during which the ovaries are relatively inactive while the female is rebuilding fat stores.

Most breeding occurs in the spring, though it may extend into midsummer, and fall mating has been observed in some locations. In Wisconsin, Vogt (1981b) reported observing mating near Madison in early April "amid the crushed ice just after the break-up of Lake Mendota." Mating takes place in the water, where males are attracted to female pheromones. Upon finding a female, the male moves to face her from the front, where he strokes her head and neck with the backs of his foreclaws. If she is receptive, she in turn strokes his forelimbs with her foreclaws. The male will periodically swim away and return until the female sinks to the bottom, where the male mounts her, grasps her carapace with his elongated foreclaws, and makes cloacal contact by bringing the vents together for copulation.

Nesting occurs from late May to mid-July, primarily in the early morning and late afternoon or sometimes at night, when females dig flask-shaped nests 3.9 to 4.3 inches (10–11 cm) deep on a south-facing slope with sparse vegetation and loamy or sandy soil. In Clay and Dickinson Counties, females were found laying eggs along the roadsides and in ditches and cornfields adjacent to wetlands 3.3 to 1,640 feet (1–500 m) from open water (Ratterman and Ackerman 1989). Females may return to the same nesting area year after year. When searching for a suitable nest site, the female may dig several test holes, and she may place her chin and neck against the ground and "nuzzle" it—possibly testing the soil texture, temperature, or moisture or smelling it (Ernst and Lovich 2009).

Hatchling western painted turtle from Davis County. Photo by Ryan D. Rasmussen.

Stromsten (1923) described in detail the western painted turtle nest digging and egg laying he observed on June 30 in Dickinson County, noting that the female dug with her hind feet, alternating among digging with one foot, urinating in the hole to soften the soil, and digging with the other foot. This sequence continued until the dirt became the "consistency of thick cream," at which time she stopped wetting it but continued digging. After about an hour and a half of digging, she began laying eggs. After each egg was deposited in the nest, she used a hind foot to push the egg to the side of the chamber. Eggs were laid at approximately 30-second intervals, with a total of 6 eggs laid in approximately 5 minutes. Once the eggs were laid, she immediately began filling in the nest, first with the dry dirt, packing it as she went, then with the wet dirt, which was "thoroughly kneaded by the knuckles, and flattened and packed by rubbing the plastron over it." Dry dirt and grass were then spread on top. Once nesting was complete, the female returned to the water; the entire process of nest digging and egg laying took over 2 hours.

Each female lays 2, possibly 3, clutches of 4 to 23 white or cream elliptical leathery eggs 1.2 inches (3 cm) long per clutch per year. Iowa clutch size reports include 5 to 13 eggs (Blanchard 1923), 6 to 13 eggs (Stromsten 1923), and an average of 12 eggs (Ratterman and Ackerman 1989). In an Illinois study on the Mississippi

River, over 30 percent of clutches had multiple paternity, and clutch sizes were larger for nests fertilized by multiple males (Pearse et al. 2002). The eggs hatch in August or September after a 62- to 80-day incubation period. Hatchlings from nests laid early in the year emerge in the fall, while those laid later in the year overwinter in the nest (Vogt 1981b). Christiansen and Gallaway (1984) reported both fall and spring emergence of western painted turtle hatchlings in Louisa County. The nearly round hatchlings have dorsal keels; large heads, eyes, and tails; and plastrons with a large dark area; they average 1 inch (2.6 cm) straight-line carapace length.

Similar to most turtles, western painted turtles lack sex chromosomes, and the sex of hatchlings is determined by the temperature of the nest—temperature-dependent sex determination. Males are produced at incubation temperatures between 70.7 to 80.6° F (21.5–27° C); females are predominantly produced at temperatures between 84.2 to 89.6° F (29–32° C).

LONGEVITY. A wild-caught adult survived in captivity an additional 14 years, 8 months, and 8 days. Wild *C. picta* have survived 30 to 40 years and, possibly, up to 60 years (Ernst and Lovich 2009).

BLANDING'S TURTLE

Emydoidea blandingii (Holbrook) 1838

Blanding's turtle from Linn County.

IOWA STATUS. Threatened; species of greatest conservation need.

HISTORICAL SUMMARY. Both the common name and the specific epithet of the Blanding's turtle honor William Blanding, a Philadelphia physician and naturalist who collected the holotype—the single specimen upon which the description and name of a new species are based—for the species around 1830. The first confirmed specimens in the state appear to be two collected in 1884 near Davenport in Scott County (National Museum of Natural History–Smithsonian Institution). The first mention of this turtle in the literature in Iowa appears to be Osborn (1892), who noted that it was common in the state. Blanchard (1923) reported a large Blanding's turtle found west of Spirit Lake in Dickinson County in 1921, which he believed was the first record of the species in Iowa (presumably he was not aware of the National Museum of Natural History specimens). Paul Errington, professor of wildlife biology at Iowa State University from 1932 until his retirement and known as one of the four great pioneers of animal ecology, collected

Blanding's turtles in Story and Hamilton Counties during the late 1930s and early 1940s (Iowa State Natural History Collection).

DESCRIPTION. Blanding's turtles have an elongated smooth and high-domed dark brown to black carapace that is commonly marked with white to yellow speckling or irregular radiating lines and a broad first vertebral scute that touches 4 marginal scutes. The number and size of the spots on the carapace increase with age; in very old turtles, the light spots often become connected, leaving a network of black marks that make older adults appear whiter than younger turtles. The marginal scutes flare only slightly near the back legs and are not serrated. The plastron varies from yellow with large black blotches at the outer posterior corner of each scute to almost totally black, and it has a movable hinge between the pectoral and the abdominal scutes. The plastron is connected to the carapace by ligaments, not bone. The top and sides of the head are gray to black with tan to yellow spots, the upper jaw is notched, and the chin, throat, and underside of the long neck are bright yellow. The legs are generally blue-gray or black and may have some yellow scales.

Hatchlings have a dark brown to black rounded and keeled carapace, a black plastron with yellow margins that lacks a hinge, a light yellow chin and neck, and a long tail. Partial albino hatchlings have occasionally been found in the state, such as one hatched from a Worth County clutch found by Jim Christiansen and Lynn

Plastron of a Blanding's turtle from Linn County.

DeVries. As the young turtles age, they develop the spotted pattern of adults and lose the carapace keel. Ross and Lovich (1992) suggested that the spotted pattern of juveniles may mimic the surface of duckweed-covered water, thereby helping to camouflage them.

Males are larger than females and have dark upper jaws, slightly concave plastrons, and thicker tails with the vent located beyond the edge of the carapace. Females have longer plastrons and higher carapaces than males, yellow upper jaws, flat plastrons, and short tails with the vent located beneath the carapace.

SIZE. Blanding's turtles generally range from 6 to 9 inches (15.3–22.9 cm) straight-line carapace length. The largest specimen in the Drake University Research Collection is a male from Sac County measuring 9.8 inches (24.8 cm) straight-line carapace length. Dodge (1956b) reported a male from Johnson County with a straight-line carapace length of 10.2 inches (25.95 cm). The maximum straight-line carapace length reported for the species is 11.2 inches (28.4 cm).

SIMILAR SPECIES. No other Iowa turtle has a high-domed dark brown to black carapace with white to yellow speckling and a bright yellow chin. Juvenile Blanding's turtles may be confused with wood turtles, as was the case with a juvenile found in 1923 on the Iowa State campus in Story County that was misidentified as a wood turtle (Palmer 1924). Reeve Bailey (1941) later correctly identified the specimen as a juvenile Blanding's. Wood turtles have a flatter carapace, lack a plastral hinge, and have yellow to yellow-orange forelimbs.

DISTRIBUTION. *E. blandingii* is found primarily from southern Ontario, Michigan, and northern Indiana westward through Wisconsin, northern Illinois, and much of Minnesota and Iowa. Scattered populations are found westward into Nebraska and from Ohio eastward through New York to Nova Scotia.

In Iowa, these turtles are found in scattered populations predominantly in the northeast half of the state and less frequently in southwest Iowa. In the mid-1940s, the species had been documented in 25 counties; at present, records exist for 60 of Iowa's 99 counties, one of which has only a pre-1945 record. Reeve Bailey (1941) noted that Blanding's turtles were "not rare in Central Iowa" in the early 1940s, and in the early 1950s Carlander and Moorman (1951a) noted that it was "probably the turtle most often appearing in back yards and towns in Iowa."

Despite its widespread distribution across the state, by the 1990s Jim (1998) reported that Blanding's turtles had clearly declined in the state due to habitat loss, and he expressed "extreme concern" that the disjunct nature of the remaining populations and the scarcity of young suggested that many of the populations might be experiencing inbreeding and might not be reproducing. In light of this, Jim and Terry (1999) surveyed Blanding's turtle populations across northern Iowa and concluded that the species had gone through a serious decline in the state and that it had been functionally extirpated in many places in northwest Iowa due to the loss of very shallow marshy areas with nearby deeper water. Areas with suitable habitat still remained in eastern Iowa, and most populations surveyed in that part of the state had young or juvenile turtles. Recent wetland restorations in north-central and northwest Iowa have helped maintain and potentially increase some populations, but this turtle remains in jeopardy throughout much of the state, as it does throughout its geographic range.

Blanding's turtle distribution in Iowa.

Blue = pre-1960
Yellow = post-1960
Green = both pre- and post-1960

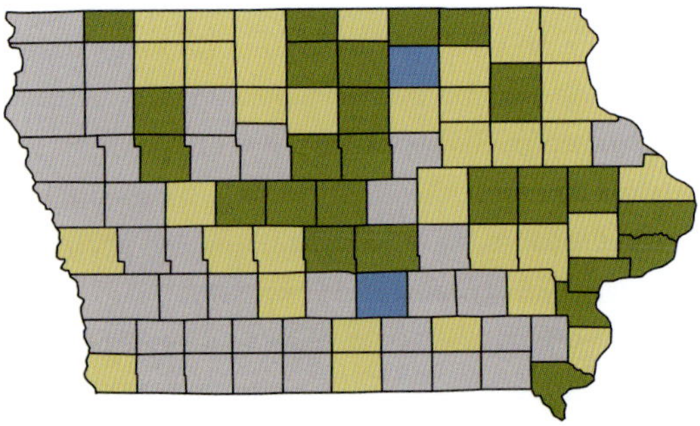

HABITAT. Blanding's turtles most commonly inhabit areas with clean, shallow, slow-moving water and abundant aquatic vegetation—*Emydoidea* is from the Greek *emydos* meaning "freshwater turtle" and *-eidos* in reference to the similarity to the genus *Emys* (pond turtles). In Iowa, they are found in lakes, ponds, marshes, fens, creeks, wet prairies, and sloughs, preferring a complex of wetlands of varying depths, but they tend to avoid marshes with a dense cover of cattails, similar to what has been observed in Wisconsin (Ross and Anderson 1990). Winter habitat includes a wetland, stream, or pond with water deep enough to prevent freezing completely to the bottom and soft substrate to allow for burying.

Among 44 Blanding's turtle records where Reeve Bailey noted habitat, 32 were on prairies or in woodland marshes or sloughs, 4 were in or near oxbows or river-edge pools, 4 were in or near drainage ditches, and 1 each was in a creek bottom, a permanent pond, a shallow gravel pit, and along a stream on the edge of the city of Des Moines in Polk County. Among northwest Iowa Blanding's turtles radio-tracked by Howing (2021), 20 percent used only wetlands throughout the year, while 75 percent used both wetland and stream habitats, and only 5 percent used only streams. In highly agricultural parts of the state, these turtles are often found in agricultural drainage ditches; in many cases, the ditches are the only remaining aquatic habitat available to them. The ditches may also serve as travel corridors between wetlands or between summer habitat and nesting areas.

Habitat preferences vary between juveniles and adults—small juveniles primarily use emergent sedge habitats and shrub hummocks, larger juveniles use sedge-water interfaces, and the largest juveniles and adults are found in open water. Jim and Reeve Bailey (1997) concluded that Iowa Blanding's turtles are primarily restricted to the vicinity of marshes and sloughs or quiet river bays where water is less than 4 feet (1.2 m) deep with dense emergent vegetation and that they tend not to remain in artificial farm ponds. In Wisconsin, ponds were preferred over marshes, though marshes likely would have been used more if open water had been more abundant, and ponds with sand bottoms and no aquatic vegetation were rarely used (Ross and Anderson 1990). During radio tracking in an urban area of Linn County where no natural wetlands remain, Terry and his coworkers found Blanding's turtles using a combination of several poor-quality habitats—the only suitable habitat remaining—including a shallow wetland that had developed in an old railroad borrow pit, a shallow marsh created by a blocked culvert under an abandoned railroad bed, a small stream, and a small artificial pond in a residential backyard. The variety of artificial habitats used illustrates the flexibility necessary for Blanding's turtles, like other species, to use lower-quality habitat where high-quality habitat no longer exists.

BEHAVIOR. These turtles are shy timid creatures that rarely, if ever, attempt to bite. They are quick to jump into the water at the first sign of danger, and if encountered on land where escape is not possible, they will pull their heads into their shells and tilt the shells in the direction of the threat. If handled, they pull their heads and limbs into their shells, occasionally utter a low hiss, and tightly close the movable part of their plastrons.

The collection/observation dates of 768 Iowa Blanding's turtles show that the species has been found every month of the year in Iowa, though they are most active from April through October. They may not be fully dormant during the winter, and at times they may move under the ice. The earliest Iowa record is one found dead on top of the ice on January 19 in Jones County. In many years, activity begins in late February and continues to increase throughout March. Sometime in April, the turtles begin moving to their summer feeding areas and can often be seen basking on land. Iowa observations peak from April through June (67% of records), during which time the turtles are moving between wetlands and females

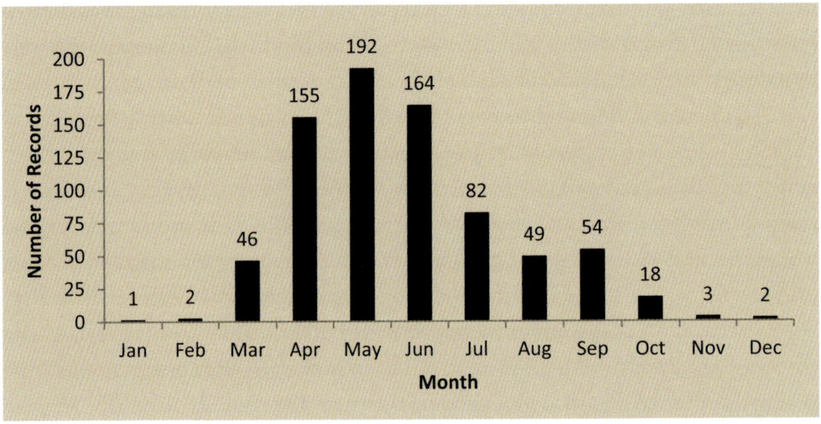

Collection/observation dates of 768 Blanding's turtles in Iowa.

are moving to and from nesting areas. After about the middle of June, observations decrease as the turtles become more aquatic and are seen less frequently on land, instead seeking shade under aquatic vegetation and limiting basking to the cooler morning hours.

Blanding's turtles are largely diurnal and may be active all day, though they tend to be more active in the morning. At night, they sleep suspended in aquatic vegetation or resting on the bottom of wetlands. This turtle frequently basks and can often be seen sunning itself on muskrat lodges, logs, stumps, other debris, and sedge hummocks and in shallow water with only the top of its carapace exposed. It also commonly leaves the water to bask on exposed mud banks near the water's edge or sometimes well away from the water, sitting in the sun with its bright yellow neck outstretched. Blanding's turtles are active thermoregulators whose daily pattern consists of a period of gradual or rapid heating in the morning, a plateau phase at midday, and rapid or gradual cooling in the late afternoon (Sajwaj and Lang 2000). They prefer a body temperature significantly lower than that of other turtles, with males favoring a lower temperature (72.5° F [22.5° C]) than females (76.6° F [24.8° C]) (Nutting and Graham 1993). In addition, their critical thermal maximum—an animal's upper thermal tolerance limit—was found to be one of the lowest among 25 turtle species examined, which may be a reason for the species' northern distribution (Hutchison et al. 1966).

Although aquatic, these turtles spend quite a bit of time on land and frequently move overland between wetlands, sometimes up to 0.6 mile (1 km). In Minnesota, males were found to move between wetlands more often but for shorter distances than females, and males used an average of 2.6 wetlands in a year, whereas females used an average of 2.1 (Piepgras and Lang 2000). Radio tracking of Blanding's turtles in Linn County by Terry and his coworkers over the course of 15 years showed that the turtles moved among 3 wetlands and a small stream throughout each year, with the turtles frequently found on land as they moved between wetlands.

During the hottest and driest parts of the summer, Blanding's turtles may estivate either in the water or on land. During this time, the turtles wait out unfavorable conditions by becoming inactive, lowering their metabolic rate, and entering a state of dormancy similar to hibernation. In Wisconsin, Blanding's turtles estivated from 0.5 to 5 days in late July and August when air temperatures were 64 to 91° F (18–33° C; Ross and Anderson 1990). During the drought year of 2012, Terry and Joshua Otten (2022b) observed terrestrial estivation in Blanding's turtles in Linn County. Five of six radio-tracked turtles estivated in terrestrial habitats during August and September, after which they moved to an adjacent wetland where they overwintered. The sixth individual estivated terrestrially for a short time in late June and early July. Individuals estivated from 7 to 43 days (mean = 24.7) 28 percent of the time they were tracked, and they were found almost entirely buried in native sand prairie, partially buried in woodland leaf litter or pine needles, or completely exposed under shrubs or trees. Several turtles used multiple locations.

As water temperatures drop in September and early October, the turtles start moving to their overwintering locations, which are generally not far from their summer habitat. Most hibernate partially buried in the mud or muck on the bottom in the deeper portions of wetlands, streams, rivers, ponds, or lakes. Terry documented Blanding's turtles in Linn County hibernating in a small artificial pond located in a residential yard. Due to the loss of other suitable winter habitat in the area, all Blanding's turtles and a large number of western painted and snapping turtles overwintered in this pond, the only waterbody with enough water to support overwintering turtles. This illustrates the need to preserve critical habitat and the ability of some individuals to use habitat that is less than optimal if needed. Blanding's turtles have metabolic adaptations that allow them to survive the low oxygen conditions that can develop due to prolonged ice cover during the winter. In areas where dissolved oxygen may be limited due to ice and snow cover, Blanding's turtles may select areas of stable and colder water temperatures. Potentially winterkills, the latest Iowa records are two dead individuals found in Greene County in December. The latest live record is from Scott County on November 27.

Hatchlings have the ability to survive freezing and may overwinter in the water or on land near the nest. Fewer reports of adult terrestrial overwintering exist. Terry and Stacey Parks (2021) reported terrestrial overwintering of an adult female Blanding's turtle in Linn County that successfully hibernated buried in the bottom of a shallow wet ditch adjacent to an active rail line.

THREATS. Although they have large heavy carapaces and hinged plastrons, Blanding's turtles, particularly the hatchlings and small juveniles, are nevertheless susceptible to predation. Small turtles are eaten by raccoons, skunks, foxes, large wading birds, and likely mink as well as the other predators that inhabit the wetlands in which they are found. Terry and Joshua Otten (2022a) documented 100 percent predation mortality among five radio-tracked juvenile Blanding's turtles in Linn County. Adults are susceptible as well. In Missouri, 31 percent of Blanding's turtles examined had foot injuries, damaged shells, or missing or damaged tails (Kofron and Schreiber 1985). Terry has seen many Blanding's turtles across the state with missing feet or legs. One in Linn County went into winter in good health and with four legs, only to emerge in the spring with just three legs and die a short time after.

Nest predation by raccoons, skunks, opossums, chipmunks, ground squirrels, shrews, and voles, along with many other animals, is a serious concern and can be devastating to small populations. During a 23-year study in Michigan, 40 to 100 percent of nests were depredated each year (mean = 78.2%; Congdon et al. 2000). In another Michigan study, most predation (84%) occurred during the first 5 days after nest completion, with 47 percent occurring within the first 24 hours (Congdon et al. 1983). Nest predation may only get worse as the number of mammalian mesopredators, such as raccoons and skunks, continues to increase.

Historically, habitat loss likely posed the most serious threat in Iowa as the draining and filling in of wetlands and sloughs for agriculture and the modification of rivers for flood control resulted in the loss of both summer and overwintering habitat. Roads also pose a serious threat, and many Blanding's turtles are killed as they move between wetlands or as females travel to nesting areas. In an effort to avoid road mortality after construction of a new 4-lane roadway in Linn County, the Iowa Department of Transportation designed and constructed an underpass under Iowa Highway 100 specifically for Blanding's turtles. The underpass, along with special fencing that directs the turtles into it, allows them to move back and forth under the road without the risk of being run over. Monitoring after construction showed that Blanding's turtles, along with western painted and snapping turtles, moved safely through the underpass between the two sides of the road.

FOOD AND FEEDING. Blanding's turtles feed mostly in the morning and evening and will forage both on land and in the water. In Missouri, they fed for 4.5 months of the year in two distinct phases—from April through the first half of July and from mid-August through mid-September prior to overwintering (Kofron and Schreiber 1985). When foraging in water, they are bottom feeders that search for food while slowly walking along the bottom and probing vegetation with their snouts; they also sit in small pools and ambush prey. The turtle rapidly expands its mouth-pharynx region to create a low-pressure area that pulls the food into its mouth—pharyngeal gulp.

This turtle is primarily carnivorous, feeding on aquatic insects, crayfish, earthworms, snails, slugs, leeches, frogs, frog eggs and tadpoles, and fish and their eggs; it also scavenges dead animals when the opportunity arises. Plant material is sometimes ingested, though plants likely do not represent primary prey items but, instead, are secondary items ingested as the turtle consumes other prey. In Michigan crustaceans made up 50 percent of the food by volume, with insects making up another 25 percent (Lagler 1943), and in Illinois snails were dominant, followed by crayfish, earthworms, and insects (Rowe 1992a). Penn (1950) reported that crayfish made up 58 percent of the diet by volume. Terry has headstarted—a conservation technique for endangered species, in which young animals are raised artificially and subsequently released into the wild—hatchlings that ate small snails, and the turtles quickly learned to catch live minnows. Examination of the stomach contents of specimens in the Drake University Research Collection found only soft material consistent with mostly digested soft-bodied prey.

REPRODUCTION. Maturation varies geographically and between populations and individuals; however, in general, females reach sexual maturity at around 14 to 20 years and males at around 14 years. Females can remain reproductive well past 50 and, unlike other older turtles studied, 55-year-old females reproduce more often than younger females (Ernst and Lovich 2009).

Mating takes place in the water and may occur in any month from April to October in Iowa. LeClere (2013) reported observing courtship in April, July, and September, and Terry (2021) observed courtship in October in Bremer County. Both sexes may mate with multiple partners, and females may store sperm. Mating is preceded by courtship involving eight different male behaviors that occur in a definite sequence and are necessary for mating to occur. Courtship begins with the male chasing the female and mounting her by climbing onto her carapace, followed by the male performing a series of tactile, visual, and likely olfactory displays, including chinning, placing his chin on the female's snout and pushing down; gulping; chin rubbing, moving his chin back and forth across the female's snout; swaying; violent swaying; and snorkeling, that lead either to termination of the attempted mating by the female or to copulation. During copulation, the male slides his tail under the female's tail and inserts his penis. Mating may last up to 30 minutes.

Among two female specimens in the Drake University Research Collection, both captured on May 19 at the same location in Louisa County, one contained enlarged ovarian follicles and the other had oviductal eggs, suggesting that ovulation in Iowa occurs in mid- to late May, with females generally laying eggs from late May through June. When ready to nest, females leave their resident wetlands, usually before sundown, and begin making their way to the nesting areas, which may be up to a mile (1.6 km) away; they often return to the same nesting area year after year. Some arrive at the nesting site the same evening, while others may take several days to reach the site. While traveling to the nesting area, females spend time under shrubs or leaf litter or in wetlands near the nesting sites.

Most nests in the Midwest are located in open grasslands with sandy or sandy loam soil, though nesting has also been observed in agricultural fields, road embankments, and residential yards, which present dangers to the females, eggs, and hatchlings. Most nesting occurs in the evening with the entire process taking up to 2.5 hours. Using their hind feet, females dig flask-shaped nests 4.7 to 7 inches (12–18 cm) deep with an egg chamber approximately 7 inches (18 cm) wide. Once the eggs are laid, the female uses her hind feet to fill in the nest. Blanding's turtles have temperature-dependent sex determination. Eggs incubated at 72.5 to 80° F (22.5–26.5° C) produce nearly all males, while those incubated at 86 to 87.8° F (30–31° C) produce all females (Ernst and Lovich 2009).

In the Midwest, clutches of 3 to 22 eggs have been reported (Vogt 1981b; Rowe 1992b; Congdon and van Loben Sels 1993; Pappas et al. 2000). Counts of enlarged ovarian follicles and corpora lutea in Iowa female museum specimens resulted in average clutch sizes of 7 eggs (range 4–12; n = 3) and 8.5 eggs (range 8–9; n = 2), respectively. A May 19 Louisa County female in the Drake University Research Collection contained 8 oviductal eggs. The average number of eggs per clutch for 13 Worth County clutches examined by Jim and Lynn DeVries was 9.1, and 3 Linn County clutches examined by Terry and Joshua Otten (2022a) averaged 11 eggs (range 9–13) averaging 1.5 inches long by 1 inch (3.9 × 2.6 cm) wide. Fertility in some Iowa populations is low. One of the clutches, with 13 eggs, examined by Jim and Lynn DeVries in Worth County was found to be entirely infertile, and a second clutch of 8 eggs contained 6 that were infertile. Similarly, fertility of the 3 Linn County clutches examined by Terry and Otten was 0 percent, 18.2 percent,

Hatching Blanding's turtle from Linn County.

Hatchling Blanding's turtle from Bremer County. Note the white egg tooth on the end of the beak.

and 55.6 percent, respectively. Although low fertility may have many causes, very few males were found in both the Worth and Linn County populations. The combination of a biased sex ratio and low fertility can have dire consequences for a population, including its extirpation.

Eggs hatch in August and September after an approximately 45- to 80-day incubation period. One Linn County egg incubated by Terry and Otten (2014) at 86° F (30° C) hatched in 48 days, while 4 others incubated at 80° F (26.7° C) hatched in 58, 59, 59, and 61 days, respectively. The hatchlings range from 1.1 to 1.5 inches (2.9–3.9 cm) straight-line carapace length; they emerge from the nest in the fall and either make their way to the water or burrow into the soil and overwinter on land.

LONGEVITY. Blanding's turtles are known to live more than 40 years. The plastron of an adult female found in Minnesota in 1988 was carved with initials and the date 1926, giving it a minimum age of 77 years (Brecke and Moriarty 1989). Terry recaptured an adult female in Linn County in 2017 that he first marked in 2003 at an estimated age of 13 to 16 years, putting its estimated age in 2017 at 27 to 30 years.

WOOD TURTLE

Glyptemys insculpta (LeConte) 1830

Wood turtle from Black Hawk County. Photo by Joshua Otten.

IOWA STATUS. Endangered; species of greatest conservation need.

HISTORICAL SUMMARY. The wood turtle has also been called the wood tortoise, striated turtle, skiddlepot, sculptured tortoise, red-legged tortoise, meadow turtle, field turtle, mud turtle, and whistling turtle. The first confirmed specimens in the state appear to be 13 collected in 1941 near Charles City in Floyd County (Iowa State Natural History Collection). The first mention of this turtle in the literature in Iowa appears to be Palmer (1924), who reported finding a wood turtle among floating algae at the edge of a small artificial lake that was part of a natural stream flowing through the Iowa State University campus in Story County. The specimen was later examined by Reeve Bailey (1941), who reported that it had been misidentified as a wood turtle and was in fact a juvenile Blanding's turtle. He went on to cite the 13 Charles City specimens as evidence of the wood turtle's presence in Iowa.

DESCRIPTION. The wood turtle's distinctive features make it one of Iowa's most striking turtles. The tan, grayish, or brown carapace has a heavily sculptured appearance resulting from the accumulation of raised growth rings that give the scutes an irregular pyramidal appearance—*Glyptemys* is from the Greek *glypt* meaning "carved" and *emys* meaning "turtle"; *insculpta* is Latin meaning "engraved, sculpted." The rough carapace has a vertebral keel and concentric grooves and ridges that radiate downward from the irregular pyramids. A pattern of black or yellow lines may be visible on the vertebral and pleural scutes with lines radiating downward from the upper corners of the scutes and intersecting the ridges. The undersides of the marginal scutes are often pigmented with black blotches in the outer posterior corners. The posterior marginal scutes are serrated and are strongly flared in juveniles and females. The carapace may be worn smooth in older individuals.

The plastron is yellow to cream, hingeless, and deeply notched posteriorly with a pattern of black blotches on each scute that is individually unique (Cowin and Cebek 2006), though the pigmentation can fade with age. The black head may be sprinkled with faint yellow dots, and the beak is strongly hooked and notched. The skin is brown with the exception of the throat, neck, tail, and underside of the forelegs, which are yellow to yellow-orange in midwestern wood turtles with the intensity of the coloration varying seasonally and by sex—males have higher-intensity orangish coloration that fades after the breeding season. Hatchlings are 1.1 to 1.6 inches (2.8–4 cm) straight-line carapace length with a rough gray or brown carapace that lacks a vertebral keel, an unmarked dark head and legs, and a long tail.

Males are larger than females, have a deeply concave plastron, a thicker tail with the vent located beyond the edge of the carapace, and prominent scales on the leading edge of the forelimbs. Females have a flat plastron and a shorter tail with the vent located beneath the carapace.

SIZE. Wood turtles generally range from 5.5 to 7.9 inches (14–20.1 cm) straight-line carapace length. The single specimen in the Drake University Research Collection is from Floyd County measuring 6.2 inches (15.8 cm) straight-line carapace length. Based on over 140 Iowa wood turtles measured by Jeff Tamplin and his students at the University of Northern Iowa (personal communication, 2022), straight-line carapace length in Iowa averages 2.2 inches (5.6 cm) for yearlings (1.5–1.75 years old), 6 inches (15.3 cm) for juveniles and subadults (3–14 years old), 7.3 inches (18.5 cm) for adult females (15 years or older), and 7.7 inches (19.5 cm) for adult males (15 years or older). The record straight-line carapace length reported for the species is 9.9 inches (25.1 cm).

Female (left) and male wood turtle plastrons from Butler County. Note the male's longer tail. Photo by Jeffrey W. Tamplin.

SIMILAR SPECIES. No other Iowa turtle has a carapace with a highly sculpted appearance and yellow to yellow-orange on the forelimbs. Juvenile Blanding's turtles may be confused with wood turtles, as was the case with a juvenile Blanding's turtle found in 1923 in Story County that was misidentified as a wood turtle (Palmer 1924) and later correctly identified as a juvenile Blanding's (Bailey 1941). Blanding's turtles have a high-domed carapace, a plastral hinge, and a bright yellow chin and neck; they lack coloration on the underside of the limbs.

DISTRIBUTION. Wood turtles are found in scattered disjunct populations from Nova Scotia and New Brunswick south to northern Virginia and west through southern Quebec, southern Ontario, northern Michigan, Wisconsin, eastern Minnesota, and a small portion of northern Iowa. Over 80 percent of the wood turtle's range is defined by the southern lobes of the Wisconsinan phase of the Pleistocene glaciation, with the major geographic areas inhabited being either the Canadian Shield or the northern portion of the Appalachian Mountains. Iowa's population represents a rarity in that the turtles inhabit the southern edge of the Pleistocene glaciation in what was once the forest-prairie ecotone—ecotones are transitional zones between two different forms of vegetation.

Archaeological evidence suggests that wood turtles once occupied a much wider range in Iowa. A partial wood turtle shell, dated to the Late Pleistocene—approximately 10,220 years before the present—was collected from the west branch of the east Nishnabotna River in Mills County in southwest Iowa (Hill 2022). Other Late Pleistocene remains indicate that wood turtles moved south ahead of the advancing ice as far as Tennessee, Georgia, and Mississippi (Hill 2022). As the ice retreated at the end of the Pleistocene, wood turtles once again reoccupied their former northern range. The Mills County specimen suggests that they also moved south and west into a previously undocumented western range, where somewhat cooler summer temperatures existed until approximately 10,200 years before the present. As the climate warmed, the summers were probably too warm and the environment too dry to support them (Hill 2022).

Today, wood turtles are limited to the Cedar River drainage in northern Iowa from Black Hawk County north to the Minnesota border, including portions of the upper Cedar River, the west fork of the Cedar River, Shell Rock River, and the Winnebago River, with the Iowa population likely extending into southern Minnesota (Otten 2017). The Iowa population, however, appears to be genetically separate from the Minnesota population (Spradling et al. 2010). In the mid-1940s, the species had been documented only from Floyd County; at present, records exist for 7 of Iowa's 99 counties. In the early 1950s, Carlander and Moorman (1951a) stated that the only location from which the wood turtle was known was near Charles City in Floyd County. In the 1970s, Jim Christiansen and Russ Burken (1978) listed it as rare and endangered and asked the public to report observations of the species. In the early 1980s, Jim (1981) suggested that the turtle was so rare that it might be extirpated from the state. Although there were new wood turtle records in the state by the late 1990s, Jim (1998) still considered the turtle to be endangered in Iowa. The wood turtle remains imperiled in Iowa as it does throughout its geographic range.

Wood turtle distribution in Iowa.
Yellow = post-1960
Green = both pre- and post-1960

HABITAT. Wood turtles reside in a variety of habitats across their range but are never found far from water. Studies have shown that Iowa wood turtles are usually found within 656 feet (200 m) of water (Tamplin 2019). The turtle is highly aquatic, inhabiting moving water such as rivers and streams as well as associated shoreline habitats. Streams with hard sand or gravel bottoms, clear water, and a moderate current are preferred; however, not surprisingly, the Iowa rivers inhabited by wood turtles are relatively turbid, though the current and dissolved oxygen concentrations are still suitable (Jeff Tamplin, personal communication, 2022). The turtles shift their habitat preferences throughout the active season, becoming increasingly terrestrial during the spring and summer and mostly aquatic in the fall as they prepare for overwintering. Studies have shown that Iowa wood turtles spend approximately 50 percent of the active season on land and 50 percent in the water (Williams 2013; Tamplin 2019). The turtle uses a wide variety of holes and depressions for shelter, including rabbit holes, hollow logs, and overhangs along gullies (Brown and Lapin 2022).

G. insculpta is an ecotonal species occupying a mosaic of forests and open-canopy and forest openings adjacent to the rivers and streams in which they forage (Otten 2017). Terrestrial habitats are occupied at different times of the year for nesting, thermoregulation, resting, and sometimes mating. Based on 1,351 observations in Black Hawk and Butler Counties, Jeff Tamplin (2019) reported that Iowa wood turtles spend 45 percent of their time in rivers, streams, and creeks, followed by emergent areas (grasses, forbs, saplings; 30%), deciduous forests (10%), grasses and willow saplings (9%), ponds, marshes, and ephemeral pools (2%), and sparsely vegetated riverbanks (2%); they also bask on logs (2%). Typically, males spend more time in the water than do females, potentially searching for mates by traveling in rivers and tributaries (Kaufmann 1992a, 1992b). During the summer, females may estivate terrestrially more than males.

BEHAVIOR. These turtles are good-natured and generally calm when handled, though some large males may bite and scratch, and other individuals may empty their bladders. Their learning ability is well documented. In a food-finding experiment using a maze, the wood turtle's learning ability was found to be similar to that of a rat (Tinklepaugh 1932). It is thought that this ability may aid in the turtle's homing skills. Wood turtles also exhibit a dominance hierarchy within and between the sexes, and dominant males will displace subordinates during copulation. Twenty-seven nonreproductive behaviors have been documented, including fifteen aggressive acts, eight defensive acts, and four neutral acts—feeding or resting close to another turtle (Kaufmann 1992b). Dominant turtles,

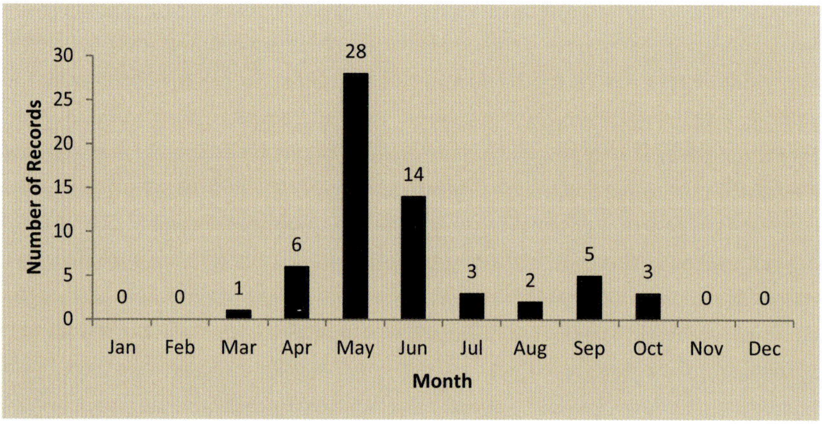

Collection/observation dates of 62 wood turtles in Iowa.

typically the oldest and largest, maintain a linear hierarchy through aggression; subordinate individuals react with submission or neutrality. Higher-ranked males have relatively greater reproductive success within the population.

Through the use of radiotelemetry, Jeff Tamplin (2019) recorded activity in Iowa from early March through mid-November, similar to the collection/observation dates of 62 other Iowa wood turtles, showing that the species has been found from March through October in the state. The earliest Iowa record is from early March in Franklin County; Tamplin (2019) has recorded activity as early as March 2 in Black Hawk and Butler Counties. Males may be active earlier in the season and remain active later than females.

Throughout the year, wood turtle activity can be divided into five periods: spring aquatic or prenesting, nesting, summer postnesting, fall aquatic, and overwintering. The spring aquatic or prenesting period—March to April—begins when water and air temperatures rise sufficiently to trigger activity. The turtles may be active on land during the warmth of the day and return to the water at night as the air temperature drops. Iowa wood turtles use terrestrial habitats nearly 75 percent of the time during this period (Williams 2013), and females spend more time basking than males in early spring to facilitate egg development.

Sometime in May or June, the nesting period, females move to their nesting sites, and both males and females stay closer to water at this time. During the postnesting summer terrestrial period—late June to early October—the turtles range relatively farther away from water in search of food compared to other times of the year, often spending several days on land before returning to the water.

If the air temperature rises above 86° F (30° C), the turtles may estivate (Dubois et al. 2009; Flanagan et al. 2013) under sticks and grass or partially buried in the soil and may not move for days or weeks at a time until the temperature drops. Tamplin (2009) tested 3-, 6-, and 12-month-old wood turtles in a temperature gradient and found that 6- and 12-month-old turtles selected 86° F (30° C) more frequently than temperatures between 59 and 80.6° F (15–27° C). While the 3-month-old turtles also oriented to warmth, some individuals were also found at lower temperatures, suggesting that temperature preference did not fully develop until 6 months.

With the arrival of autumn comes the fall aquatic period—early October to mid-November—during which the turtles divide their time equally between terrestrial and aquatic habitats as they prepare for overwintering (Williams 2013). During this time, they may make long-distance movements to congregate around hibernacula. Specific locations and habitat preferences—terrestrial or aquatic—are determined by air temperature, water temperature, and the availability of sunlight.

As fall progresses, activity decreases as water temperatures fall below 43° F (6° C; Jones and Willey 2021), at which time the turtles begin preparing for overwintering by traveling, sometimes long distances, to congregate at hibernacula. Overwintering—mid-November to early March—occurs in streams and rivers with flowing water and high levels of dissolved oxygen that typically do not completely freeze during the winter (Greaves and Litzgus 2007). Overwintering sites include deep pools either on top of the substrate or buried in the mud or sand, undercut banks, overhanging roots, logs, or beaver or muskrat burrows. In Iowa, overwintering locations were in streams with a slow current and water approximately 3.3 to 4.9 feet (1–1.5 m) deep and 3.3 to 9.8 feet (1–3 m) from the bank (Otten et al. 2021). Turtles were observed resting mostly exposed on the bottom on sandy flat substrate, wedged under submerged logs and root balls, or in logjams and accumulations of flood debris in undercut banks. Activity is limited while overwintering, though movement has been observed (Jeff Tamplin, personal observation, in Otten 2017). The latest Iowa record is October 12 in Bremer County, though Tamplin (2019) has recorded activity into the second week of November.

With the exception of nesting females, wood turtles are primarily diurnal, and much of their activity is explained by thermoregulation as determined by a combination of temperature, humidity, and time of year. They spend much of the day basking on partially submerged logs or streambanks or in woodland openings. Road ditches are also used in Iowa (Otten et al. 2021). Basking is concentrated in the late morning and late afternoon on sunny days. Individual turtles return to

Basking wood turtle from Butler County. Photo by Ryan D. Rasmussen.

the same basking sites from year to year, sometimes traveling great distances to get to an optimal site. Otten et al. (2021) found that some Iowa individuals moved more than 0.3 mile (0.5 km) to suitable basking sites.

Joshua Otten and his coworkers (2021) conducted a two-year radiotelemetry study looking at wood turtle movement, home range size, and site fidelity in Iowa. They found that home range size of adult male and female turtles differed significantly, with male average home ranges being nearly three times those of females. Females were found to use 4.2 to 57.3 acres (1.7–23.2 ha) during the course of a year, while males used 8.4 to 231.5 acres (3.4–93.7 ha). Stream home range length differed significantly between males and females as well, with males covering 0 to 2 miles (0–3.3 km) throughout a year and females moving 0.1 to 1.4 miles (0.2–2.3 km) per year. Both sexes showed a high degree of site fidelity to annual home ranges but not to specific overwintering locations. Williams (2013) estimated wood turtle density in a 1,159-acre (469-ha) area of river and floodplain in Iowa to be 0.23 turtle per acre (0.1/ha), much lower than studies from other states.

THREATS. Wood turtles have a keen sense of hearing, responding quickly to low-frequency sounds (less than 500 hertz), which helps them avoid predators (Ernst and Lovich 2009). Nevertheless, they are still susceptible to predation. As with other turtles, raccoons, skunks, mink, and other larger mammals eat wood turtle eggs and can be significant nest predators. Between 2003 and 2006, approximately 36 percent of wood turtle nests at Iowa study sites were destroyed by predators prior to hatching (Spradling et al. 2010). Fast-growing plants can also result in reduced hatching success or complete nest failure by encapsulating the eggs in roots (Behler and Castellano 2005). Small turtles are eaten by predatory fish, snapping turtles, crows, wading birds, and numerous mammals. Adult turtles can be killed or injured by raccoons, badgers, skunks, river otters, coyotes, and feral dogs and cats. In Michigan, 9.7 percent of marked wood turtles were missing one leg and over 2 percent were missing two limbs (Harding and Bloomer 1979), and similar injuries have been observed in Iowa (Tamplin 2019).

From 2003 to 2019, Jeff Tamplin (2019) and his students observed 43 deaths (29.9%) among 144 marked wood turtles in Black Hawk and Butler Counties. Causes of death included predation, unknown or natural causes, flooding, infection or indirect predation, road mortality, and poaching—the shell of the poached turtle was found with a large bullet hole in the carapace. Mortality rates did not appear to differ between the sexes, and mortality events usually occurred in terrestrial habitats (83.7% of all deaths) and peaked during late spring and early summer. Most of the observed mortalities occurred in May (16.3%) and June (27.9%). Despite the fact that the turtles were 75 percent aquatic during the fall aquatic period, late fall represented another high-risk period for predation (23.2% of all observed deaths). Most fall deaths appeared to be linked to terrestrial predator attacks during warm periods that promoted terrestrial basking.

Iowa is already seeing the effects of climate change in the form of increased frequency and intensity of extreme rain events. In areas where wood turtles occur in the state, flooding occurred in 13 of 20 years between 1989 and 2008, with 6 years involving major or moderate flooding during the peak nesting times of June and July (Otten 2017). Flooding appears to be the primary cause of nest failure in Iowa, with nearly 64 percent of nests destroyed by flooding during a 2003 to 2006 study, during which all known nesting sites in Iowa experienced flood events throughout the incubation period (Spradling et al. 2010). Row-crop production and urban development in streamside habitats can also harm wood turtle populations. While wood turtles mainly occupy floodplain habitats, adjacent natural upland habitats are also critical, but these have become fragmented and altered by agriculture and development—land uses inhospitable to this habitat specialist (Tamplin 2019).

Because of their attractiveness and temperament, wood turtles have been popular in the pet trade, and collecting them as pets has damaged some local populations. Although collection is prohibited in most areas today, illegal collection does still occur, as evidenced by 10 wood turtles found during a search by Iowa Department of Natural Resources conservation officers in the late 1990s at the eastern Iowa residence of someone subsequently prosecuted for selling reptiles illegally.

FOOD AND FEEDING. Along with thermoregulation, food availability explains much of the wood turtle's activity (Tamplin 2019). Wood turtles are omnivores that consume mushrooms, algae, grasses, berries, and other plants along with a variety of invertebrates such as beetles and other insects, millipedes, earthworms, slugs, and tadpoles as well as vertebrate carrion. In Iowa, the turtles consumed in order of frequency of observation grasses, slugs, violet leaves, prairie ragwort leaves—a highly toxic plant known to kill fish, lizards, and livestock—black raspberries, earthworms, snails, and dandelion leaves (Tamplin 2006). Green leaves—mostly forbs like violets and goosefoots—and grasses are the most common food items, with slugs being the most commonly consumed invertebrate in Iowa (Jeff Tamplin, personal communication, 2022). Tamplin and his coworkers (2009) reported consumption of a ring-necked snake and the shed skin of a garter snake in West Virginia, but it is unknown if the ring-necked snake was already dead when eaten or was mistaken for an earthworm. Wood turtles occasionally eat the eggs of their own species; Tamplin has observed several cases of Iowa wood turtles eating the eggs of other wood turtles in captivity (Jones and Willey 2021). The turtles may also eat a poisonous mushroom, the fly agaric, which contains a toxin that is stored in the turtle's tissues. While the turtle is unharmed by the toxin, consuming the turtle can be fatal to a would-be predator.

When feeding on land, the turtle slowly pokes its head into clumps of vegetation around the base of bushes, logs, or debris piles looking for food. In the water, its behavior is much the same as it investigates clumps of algae, emergent vegetation, or the entrances of muskrat burrows. Carlander and Moorman (1951a) noted that wood turtles use their forelegs while feeding more than other turtles and are reported to hold strawberries with their front legs while biting and tearing them apart.

Several studies have documented an unusual feeding behavior known as worm stomping, during which the turtle stomps the ground with its front feet and plastron to draw worms to the surface where they can be captured and eaten. It is well known that earthworms are drawn to the surface during rain events, and it is commonly believed that they quickly head for the surface during rain to avoid drowning in saturated soil. It has long been thought that worm stomping mimics the vibration of raindrops hitting the soil, which drives the worms to the surface.

However, research by Catania (2008) has shown that it is not rain the earthworms are fleeing but predators—moles. In the 1800s, Charles Darwin too believed that earthworms were escaping the vibrations of moles in the soil, but he was unable to prove it. Catania's research supports the hypothesis that earthworms have a stereotyped escape response to foraging moles consisting of rapidly exiting the soil, which both prevents pursuit by the mole and allows movement away from the mole and subsequent burrowing elsewhere. Unfortunately for the earthworm in the case of worm stomping, another predator is waiting at the surface, where the wood turtle takes advantage of what has been called the rare enemy effect—a scenario in which a predator catches its prey by exploiting a response that is usually a good strategy.

REPRODUCTION. Maturation occurs between 11 and 20 years and varies both geographically and between populations and individuals (Jones and Willey 2021). In general, females reach sexual maturity at around 5.3 to 6.3 inches (13.4–16 cm) straight-line carapace length and males at approximately 5.6 to 7 inches (14.1–18 cm) straight-line carapace length.

Mating usually occurs in the spring and fall but, depending on geography, can occur at any time of the year. Most courtship and copulation occur in the water at depths of 4 to 47 inches (10–120 cm) and water temperatures from 49 to 68° F (9.5–20° C), though terrestrial courtship and mating have also been observed. The male usually initiates courtship by nosing the female's shell, head, tail, and legs, and if she moves away, he will pursue her. Pairs of turtles may display to each other for over an hour with head bobs and head sways prior to mounting.

While courting, it has been suggested that the turtles may produce a very subdued whistle (Pope 1939), and Carlander and Moorman (1951a) noted that wood turtles are known as whistling turtles because of the teakettle-like whistle they give during the courting season. However, Harding and Bloomer (1979) mentioned that they had never heard a wood turtle make any noise that could be interpreted as intentional, but they did have an individual with an apparent respiratory problem that whistled when it exhaled.

Mounting may last from 1 to 12 hours as the male clasps the rear of the female's carapace, shakes her from side to side, rubs his plastron on her carapace, and bites both her head and her plastron. Males will also audibly thump the female's plastron for 5 minutes to almost 2 hours. During copulation, which may last up to 2 hours, the female may crawl to shallow water so she can breathe. Females may mate with more than one male in a season and can store sperm over the winter.

Sometime in late May or June, females make their way to their nesting areas,

Hatchling wood turtles from Butler County. Photo by Jeffrey W. Tamplin.

at times covering themselves with sand as they leave the water, presumably as a form of camouflage (Harding and Bloomer 1979). They often return to the same area to nest year after year. In Quebec, 95 percent of females returned to the same nest site (Walde et al. 2007). In Iowa, nesting dates are almost always during the first two weeks of June, with warm or cold springs seeming to have little effect on these dates (Jeff Tamplin, personal communication, 2023). Riverine sandbars and sandy riparian habitats with good solar exposure are preferred; rocky locations, waterlogged soils, flood-prone areas, or thickly vegetated areas are avoided.

Most nesting occurs during daylight hours with peaks of activity in the morning and evening, and the entire process takes up to 3 hours. Using her hind feet, the female digs a hole into which the eggs are laid. Once all the eggs are laid, she uses her hind feet to fill in the hole by scraping its sides with her claws and pushing dirt from the surface, tamping it as she goes to pack the soil. She may also pound the dirt with the rear of her plastron to match the surrounding ground.

Females lay a single clutch of 3 to 20 thin-shelled leathery eggs per year; the eggs are white, oval, and 1.5 inches (3.7 cm) long. Clutch size varies geographically but is positively correlated with female size and latitude; occasionally, a female may split a clutch between 2 nests. In Wisconsin, Vogt (1981b) reported that 8 eggs per clutch were most common; however, an average of 11 eggs per clutch was documented in northern Wisconsin (Brown and Lapin 2022). In Michigan, an average of 10 eggs per clutch was reported (Harding and Bloomer 1979). In Iowa, Tamplin found an average of 10 eggs per clutch (range 6–13) for 15 wood turtle nests recorded in Black Hawk and Butler Counties from 2004 to 2019 (Jones and Willey 2021), and LeClere (2013) reported a clutch of 11 eggs laid on June 15 by a Cerro Gordo County female. Unlike most Iowa turtles, wood turtles do not have temperature-dependent sex determination; instead, sex of the hatchlings is genetically determined (Ernst and Lovich 2009). Eggs hatch in August and September after an approximately 40- to 65-day incubation period. The Cerro Gordo County clutch reported by LeClere hatched on August 8 after being incubated at 81° F (27.2° C). The hatchlings range from 1.1 to 1.6 inches (2.8–4 cm) straight-line carapace length and emerge from the nest in the fall.

LONGEVITY. Several studies document wood turtles living at least 50 years or more in the wild, and an individual in the London Zoological Gardens lived 100 years in captivity (Jones and Willey 2021).

NORTHERN MAP TURTLE

Graptemys geographica (Lesueur) 1817

Northern map turtle from Mitchell County. Photo by Chelsea Ewen Rowcliffe.

IOWA STATUS. Protected; species of greatest conservation need.

HISTORICAL SUMMARY. The northern map turtle has also been called Lesueur's terrapin, geographic terrapin, geographic tortoise, map tortoise, geographic turtle, hackleback, and common map turtle. "Common" is no longer used to avoid any misinterpretation that suggests abundance rather than the species' widespread geographic range. The first confirmed specimen in the state appears to be in 1930 from the Mississippi River near Keokuk in Lee County (National Museum of Natural History–Smithsonian Institution). However, the *History of Fremont County, Iowa* (Anonymous 1881) listed the northern map turtle as occurring in the county. Also, Osborn (1892) reported one from West Liberty in Muscatine County with the note, "I have seen one specimen which was shown me by Prof. S. E. Meek at Coe College, Cedar Rapids," and Somes (1911) stated that the species was found in Iowa but did not reference a specimen.

Plastron of a northern map turtle from Mitchell County. Photo by Chelsea Ewen Rowcliffe.

DESCRIPTION. Northern map turtles have a low flattened olive or dark brown carapace with a low vertebral keel and a network of yellow or orange lines that resemble a topographic map, giving rise to the common name. Conant (1975) described this pattern as resembling "an intricate system of canals or waterways laid out on a chart." The scientific name as well is derived from this pattern—*Graptemys* is from the Greek *graptos* meaning "inscribed" or "engraved" and *emys* meaning "turtle," possibly in reference to the markings on the carapace that resemble a map; *geographica* is from the Greek *geo* meaning "world," *graph* meaning "map," and *-ikos* meaning "belongs to," again in reference to the map-like carapace markings. The posterior marginal scutes are serrated, and a dark blotch is sometimes found on each scute of the carapace of older adults. The smooth hingeless plastron is cream to yellow in adults. The ventral side of the marginal scutes is yellow with dark circular markings, and the plastral bridge has longitudinal dark lines. The head, neck, and legs are olive to dark brown with yellow to greenish stripes. Hatchlings are patterned like adults, including yellow spots behind the eyes; however, they are brighter in color and have prominent vertebral keels and dark coloration along the seams of the plastron.

The species is distinguished by its head markings. A small green to yellow triangular mark that is not connected to any of the neck or head stripes is located just behind the eye. The anterior end of one head stripe often turns upward toward the tympanum, though some head stripes reach the eye. The lower jaw has longitudinal yellow stripes.

Northern map turtles exhibit pronounced sexual dimorphism, with males being smaller than females with oval carapaces, low vertebral spines, smaller heads, and longer and thicker tails with the vent located beyond the edge of the carapace, whereas females are significantly larger with broader heads, carapaces that are more rounded, no vertebral spines, and shorter tails with the vent located beneath the carapace. The coloration of the head behind the eyes of adult males is brighter and more saturated in yellow pigment compared to adult females but, interestingly, no color differences occur between adult males and juvenile females (Bulté et al. 2013).

SIZE. Female northern map turtles generally range from 7 to 10.7 inches (18–27.3 cm) straight-line carapace length; males generally range from 3.5 to 6.3 inches (9–16 cm) straight-line carapace length. The largest specimen in the Drake University Research Collection is a female from Muscatine County measuring 9.2 inches (23.3 cm) straight-line carapace length. The record straight-line carapace length reported for the species is 10.7 inches (27.3 cm).

SIMILAR SPECIES. Iowa's three *Graptemys* species may be found together and can be easily confused. Both the northern false map turtle and the Ouachita map turtle have well-developed vertebral keels with prominent spines and more prominently serrated posterior marginal scutes. Northern false map turtles have a narrow yellow mark shaped like a backward L behind the eyes and no spots on the lower jaw. Ouachita map turtles have a large yellow square, rectangular, or oval mark behind the eyes, a large spot under the eyes, and another spot on the lower jaw directly below the eyes.

Adding to the confusion, all three of Iowa's *Graptemys* species are known to hybridize with each other. Studies quantifying molecular diversity in *Graptemys* indicate that interspecies mating occurs (Freedberg and Myers 2012; Mitchell et al. 2016), and Vogt (1978) documented hybridization in the Mississippi River in Wisconsin. Different feeding niches and courtship behavior may be maintaining species segregation; nevertheless, individuals with characteristics intermediate among the three species are found in Iowa.

DISTRIBUTION. The northern map turtle has the widest range among the North American map turtles, extending from southern Quebec and northwest Vermont through the St. Lawrence River drainage to the Great Lakes, west through

Michigan, Wisconsin, and eastern Minnesota, south along the Mississippi River drainage through eastern Iowa, Illinois, Missouri, Arkansas, and Louisiana, west to eastern Oklahoma and Kansas, and east to the Adirondack and Appalachian Mountains.

In Iowa, *G. geographica* is found in the eastern half of the state associated with the Mississippi River and some of its tributaries. In the mid-1940s, the species had been documented in 5 counties; at present, records exist for 23 of Iowa's 99 counties. Otten (2020) added 8 new county records along the Cedar River in Mitchell, Bremer, and Black Hawk Counties, the Wapsipinicon River in Buchanan and Jones Counties, the Iowa River in Marshall and Hardin Counties, and the Mississippi River in Des Moines County.

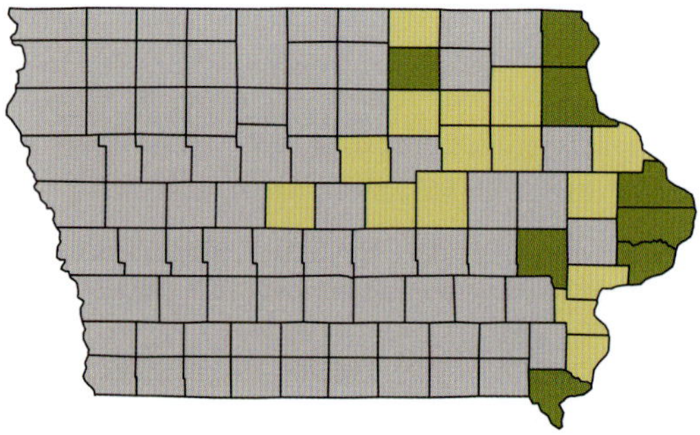

Northern map turtle distribution in Iowa.
Yellow = post-1960
Green = both pre- and post-1960

In the early 1980s, based on sampling conducted between 1970 and 1980, Jim Christiansen (1981) suggested that northern map turtle populations in Iowa were declining. In the early 1990s, Terry (1993) reported that no specimens had been taken from the lower Wapsipinicon, lower Skunk, or lower Des Moines Rivers. Since then, two records have been documented in the lower Wapsipinicon River in Jones and Scott Counties, but there are still no records in the lower Skunk or lower Des Moines Rivers, suggesting that northern map turtles are still quite rare in the larger downstream reaches of these rivers. An explanation for this lack of downstream turtles comes from studies by Terry and Jim (1996), who reported that these turtles were among those most affected by stream modification in the state and had been eliminated or severely reduced in number in the most heavily modified rivers.

HABITAT. Northern map turtles are primarily a riverine species found in medium to large rivers in Iowa, though they also use adjacent sloughs and oxbows. They prefer areas with slow-moving water, numerous basking sites, a gravel or rocky bottom, and some aquatic vegetation; however, no significant relationship was found between northern map turtle abundance and any measured river habitat characteristic in the St. Croix River in Minnesota and Wisconsin (DonnerWright et al. 1999). In a Pennsylvania river, 52.9 percent of captures were in a deep and slow area, 17.8 percent were in a shallow and slow area, 2.3 percent were in riffles, and 27 percent were in areas where depth and water velocity were variable (Pluto and Bellis 1986). In the St. Lawrence River in Ontario, the turtles avoided water more than 6.6 feet (2 m) deep and selected home ranges in water less than 3.3 feet (1 m) deep (Carrière and Blouin-Demers 2010).

Closer to Iowa, most northern map turtles collected as a by-product of fisheries monitoring in the upper and middle Mississippi River from 1992 to 1995 were captured in backwater areas (Tucker, Gritters et al. 2008). Similarly, Anderson et al. (2002) reported that northern map turtles were more commonly found in backwaters and sloughs with permanent water, slow currents, and abundant basking sites in a reach of the upper Mississippi River across from Keokuk, Iowa.

Habitat preference differs among groups, with females using deeper water close to shore and males choosing shallower areas with emergent and floating vegetation (Carrière and Blouin-Demers 2010); larger turtles use deeper water and faster currents than smaller turtles (Pluto and Bellis 1986). Land use adjacent to the river may also affect habitat use. Ontario northern map turtles showed a preference for natural shorelines over developed shorelines (Carrière and Blouin-Demers 2010), and river occupancy was more associated with agricultural habitat than with grassland-shrub habitat in north-central Indiana (Rizkalla and Swihart 2006).

BEHAVIOR. These shy turtles are highly aquatic and rarely found far from the water. If approached while basking, even from a distance, they will quickly slip or dive into the water, sometimes from several feet above the water. Vogt (1981b) considered a large female northern map turtle to be an "ominous threat to a canoeist's head" and suggested installing signs along rivers where the turtles occur that read, "Watch out for falling turtles." Neil was once challenged by his herpetology professor to capture a basking map turtle; spying one sitting on a rock down a forested slope, he charged down the slope, ran through the shallow stream, and almost had his hands on the turtle before it slipped off and swam away. If handled, some turtles may attempt to bite or may empty their bladders, but most will simply pull into their shells and wait for the threat to go away.

The collection/observation dates of 110 Iowa northern map turtles show that the species has been found from March to October in Iowa, though they are likely primarily active from late April through September. Carlander and Moorman

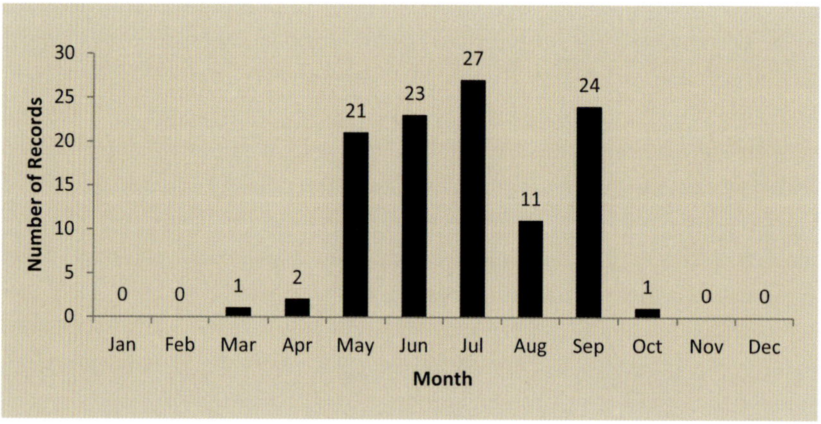

Collection/observation dates of 110 northern map turtles in Iowa.

(1951b) noted that northern map turtles apparently do not hibernate as long as most other species and are usually the first turtles to appear in the spring and the last to disappear in the fall. The earliest Iowa record is one photographed basking on a log in the Mississippi River in Scott County on March 18.

The turtles are almost entirely diurnal, spending much of the day searching for food or basking on partially submerged logs, rocks, or other debris, tree branches overhanging the water, or sandbars not connected to the shoreline. They may also bask floating on the surface of the water if the water is warm enough—aquatic basking. Foraging may occur in the morning or late afternoon, though most activity is from midmorning to noon. When not foraging, the turtles are basking, sometimes many hours each day, either singly or in large groups and often with other turtles such as northern false map turtles, Ouachita map turtles, western painted turtles, and eastern spiny softshells. Females tend to bask longer than males, particularly in spring and early summer as eggs are developing.

Otten et al. (2024) used light-level geolocators on 17 Michigan northern map turtles and found that the number of times spent moving between water and basking increased for females during late May and early June but remained relatively constant for males. For both sexes, time spent basking peaked in mid-May and then decreased for the remainder of the active season, except for an increase in late June and early July. With awareness of the complexity of thermoregulation in aquatic ectotherms, the researchers noted that time spent in aerial basking is a complex interplay of air and water temperatures in conjunction with temperature maintenance relative to physiological requirements.

Basking is necessary for the turtles to reach their preferred temperature, around 82° F (27.7° C; Nutting and Graham 1993), and it can raise their body temperature significantly above the water temperature. In addition to thermoregulation, basking can help reduce the amount of algae growing on the carapace, assist with digestion, and provide an opportunity for birds to assist in cleaning leeches from the turtles' bodies (Vogt 1979; Vogt et al. 2018). Because basking is so important to daily activities, it could be assumed that the northern map turtle is constrained by a narrow range of environmental temperatures. However, both adults and hatchlings have good locomotor abilities over a wide range of body temperatures from 66.2 to 86° F (19–30° C; Ben-Ezra et al. 2008), another adaptation allowing for their wide north-south distribution. At night, the turtle sleeps on the bottom of the river or underwater on submerged logs (Vogt 1981b).

Throughout the active season, northern map turtles may make extensive movements, with those in rivers occupying linear home ranges averaging 1 to 3 miles (1.6–4.8 km) in length encompassing areas both upstream and downstream of their overwintering sites. Two annual movement patterns were observed in a Pennsylvania river (Pluto and Bellis 1988). Beginning in late spring or early summer, adult males moved either upstream or downstream of the deep slow area of the river where they overwintered to more shallow and varied sections where they established summer ranges for around two months. Beginning in August or

Northern map turtle aquatic basking in Johnson County. Photo by Dean Hester.

September, the males once again moved back to the deep slow area. In the second pattern, both males and females stayed within a relatively small area focused around major basking sites. The researchers felt that competition for food was a primary factor inducing these movements. Similarly, Vogt (1981a) suggested that food availability rather than habitat might limit use of an area.

As fall progresses, the turtles begin congregating at their overwintering locations, likely sometime in September or early October in Iowa. The latest Iowa record is one observed basking in the Dubuque Ice Harbor in Dubuque County on October 15. Northern map turtles generally overwinter unburied on or near the bottom of deep well-oxygenated pools where the water is deep enough to prevent freezing to the bottom. In Wisconsin, muskrat and beaver burrows along with the area behind logs and rock piles were used as overwintering sites (Vogt 1981b). In Missouri, overwintering has been observed in rock crevices, between large rocks, inside large root balls, and under large submerged logs on the bottom of streams (Briggler and Johnson 2021).

The turtles may overwinter singly or clustered in small or large groups, often piled on top of each other. Groups of up to a hundred individuals have been reported (Graham and Graham 1992). They may remain active during the winter and have been observed swimming under the ice (Vogt et al. 2018), and they may bask on warm and sunny winter days (Johnson 2000). Carlander and Moorman (1951b) noted that they have even been seen walking over the lake bottom under the ice in midwinter. While they are submerged during the winter, the turtles extract oxygen from the water and carbon dioxide from the blood through a combination of cloacal and buccopharyngeal respiration.

THREATS. Raccoons, skunks, opossums, foxes, river otters, and coyotes all eat *G. geographica* eggs and can be significant nest predators. A three-year study in Wisconsin documented a 90 percent map turtle—northern map turtles and Ouachita map turtles—nest depredation rate; the median time between egg deposition and nest depredation was 12.5 hours, and 74 percent of the nests were depredated within 48 hours (Geller 2012). Birds such as common grackles, American crows, and red-winged blackbirds prey on the hatchlings as they make their way from the nest to the water, and once they are in the water herons and large predatory fish, such as largemouth bass and northern pike, eat them (Vogt 2022b). Adults, due to their highly aquatic lifestyle and wariness, are likely not commonly depredated, with nesting females being the exception. At a nesting site in Minnesota, the remains of 13 adult female northern map turtles were found within a

5-day period along with bones of other turtles (Cochran 1987). Opossums, raccoons, skunks, and coyotes were considered to be the likely predators.

Parasites may affect the health of some northern map turtles, which are hosts to a number of parasites, including leeches, trematode flukes, and roundworms. Four coccidian parasites—single-celled endoparasites—were found in the Iowa northern map turtles examined by Wacha and Christiansen (1976). Larvae of a sarcophagid fly—a flesh fly—were observed attacking Wisconsin hatchlings that were still in the nest (Vogt 1981c).

Humans pose a significant threat to northern map turtles through habitat loss resulting from the modification of riverine habitats. In the 1990s, Terry (1993) studied the effect of stream modification on riverine turtles in Iowa. The work showed that activities such as channelization, riprapping of streambanks, and impounding of rivers result in a reduction or a loss of diversity of both the types of habitats and the turtle species in riverine systems. As the habitats become more homogeneous and less complex, rare species or those with specialized habitat requirements are lost. Northern map turtles are very vulnerable to river modification. Straightening, deepening, widening, and impounding rivers, along with the removal of backwaters and oxbows, can all make rivers less suitable for these turtles.

Commercial harvesting of map turtles was allowed in Iowa until 2006 and may have posed a threat to some populations in the state in the past. Although the harvest of map turtles was never as intense as it was for snapping turtles or softshell turtles, based on Iowa Department of Natural Resources records between 1987 and 2005—the last year that map turtle harvest was allowed—commercial turtle trappers harvested an average of 325 pounds (147 kg) with a range of 0 to 2,054 pounds (0–932 kg) of map turtles per year in the state. Harvesters were not required to distinguish among map turtle species; therefore, it is not possible to know exactly how many northern map turtles were taken. However, the reported average does illustrate the magnitude of the catch.

Water pollution, invasive mussels, and other factors affecting the turtles' primary food sources also have to be considered threats. White and Moll (1992) reported that northern map turtles in the Illinois River ate snails when they were abundant in that river in the early 1900s but shifted to an almost totally insectivorous diet as siltation began to eliminate most mollusks in the 1940s.

FOOD AND FEEDING. Northern map turtles are carnivores that specialize in eating mollusks, though aquatic insects, crayfish, and fish carrion are also consumed. They swim or crawl along the bottom in search of prey, probing vegetation or sometimes turning over small rocks to get at the snails or small mussels underneath. When mollusks are found, the turtle grabs whatever soft parts are sticking out of the shell and bites them off, leaving the shell behind (Newman 1906). Failing this, it crushes the shell with its jaws, then uses its claws to pick through the crushed parts.

In Wisconsin, these turtles feed primarily on mollusks (81% of stomachs examined), fish carrion, and the larvae of caddis flies, mayflies, and damselflies (Vogt 1981a). The small amount of plant material ingested was thought to have been incidentally consumed during the capture of other prey. A Pennsylvania population increased its reliance on mollusks after the introduction and subsequent population explosion of invasive zebra and quagga mussels (Lindeman 2006). In Missouri, small snails were most commonly eaten (94.1% by volume) followed by crayfish and aquatic insects (White and Moll 1992). Other authors have also reported crayfish making up a large part of the diet (Lagler 1943; Penn 1950). Because their larger heads are adapted for crushing shells and exoskeletons, females feed primarily on mollusks while males feed more often on aquatic insects. In a Maryland river, virtually no overlap in the diets of the two sexes was found, and the researchers suggested that head size sexual dimorphism implied that two different food sources were required to maintain a stable population (Richards-Dimitrie et al. 2013).

Northern map turtles are often sympatric—occupying the same habitats—with Iowa's other map turtles. Competition among the species is reduced because each species uses the food sources in its environment differently in a way that helps them to coexist—niche segregation. Northern map turtles are carnivores specializing in mollusks, Ouachita map turtles are omnivores specializing in surface feeding, and northern false map turtles are generalist omnivores that overlap the other two species both in diet and in feeding strategies (Vogt 1981a).

REPRODUCTION. Information regarding northern map turtle reproduction comes primarily from the superb field studies of Vogt (1980, 2022b) in Wisconsin, which have been used for the following except as noted. Males reach sexual maturity at around 3 to 4 years, while females mature later at around 12 to 13 years.

Females overwinter with a full set of vitellogenic follicles that are ovulated in the spring after they emerge from overwintering, and they are able to store sperm over the winter. Mating takes place in the water in spring as well as in fall while the turtles are gathered at their overwintering locations. The courtship display involves the male approaching the female from the front, making snout-to-snout contact and then bobbing his head up and down rapidly, though he does not drum his foreclaws against her ocular regions as do other species of map turtles. When the female is motionless, the male moves into position and attempts to mount by making cloacal contact, hooking his tail around hers to maintain his balance, and bringing the vents together.

After leaving their overwintering sites, females head to the nesting areas to spend several weeks basking as the eggs form shells. Females show strong nest site fidelity, returning to the same sites year after year. Nesting occurs from June to mid-July; the females dig flask-shaped nests averaging 4.7 inches (12 cm) deep in soft sand or soil, preferably in full sunshine. Along the Mississippi, lower Cedar, and lower Iowa Rivers, nests are often located on the sandy levees that border the rivers. Most nesting occurs early in the day during overcast, humid, and warm conditions. Two to three clutches are produced annually in Wisconsin, southern Missouri, and Pennsylvania (Vogt et al. 2018). Iowa turtles likely also lay more than one clutch. Of two Muscatine County females in the Drake University Research Collection found on June 20 and 28, one contained 6 enlarged ovarian follicles, and the other contained 7 enlarged ovarian follicles and 1 oviductal egg, respectively, suggesting that these two females would have laid eggs in late July. It is unknown whether these would have been their first or second clutches of the year.

Females lay from 6 to 20 dull white ellipsoidal eggs per clutch that are 1.3 to 1.4 inches (3.2–3.5 cm) long, soft, and easily dented. In the Midwest, clutches of

Juvenile northern map turtle from Mitchell County. Photo by Chelsea Ewen Rowcliffe.

10 to 16 eggs (Illinois, Cahn 1937), 9 to 20 eggs (Wisconsin, Vogt 2022b), and 6 to 15 eggs (Missouri, White and Moll 1991) have been reported. Counts of enlarged ovarian follicles and shelled oviductal eggs in the two Muscatine County females in the Drake University Research Collection resulted in an average clutch size of 7 eggs (range 6–8). Ryan and Lindeman (2007) studied the relationship among egg size, clutch size, and female body size in northern map turtles. Their results suggest that the species employs an interesting reproductive strategy whereby females produce "the greatest number of the largest eggs that they are physically capable of laying."

The eggs hatch from August to September, with both fall and spring emergence from the nest reported. Vogt (1980) found no evidence of overwintering in the nest in Wisconsin. However, Christiansen and Gallaway (1984) reported spring emergence in Louisa County, and Baker et al. (2003) reported that hatchlings routinely overwintered in their natal nests. For those that do overwinter in the nest, freeze tolerance is poor in this species, and hatchling survival under prolonged subzero temperatures is attributed to their ability to supercool in the nest—to allow their body temperature to drop below freezing without the formation of ice crystals (Baker et al. 2003; Lindeman 2013). The nearly round hatchlings have a prominent dorsal keel and average 1 to 1.3 inches (2.6–3.3 cm) straight-line carapace length.

As with most turtles, northern map turtles lack sex chromosomes, and the sex of hatchlings is determined by the temperature of the nest—temperature-dependent sex determination. In the lab, eggs incubated at 77° F (25° C) produced only males, and those incubated at 87° F (30.5° C) produced only females (Bull and Vogt 1979). In the wild, shaded nests produced only males while those in the sun produced only females (Bull and Vogt 1979).

LONGEVITY. A wild-caught adult survived in captivity an additional 18 years and 21 days, and survival in the wild of over 20 years has been reported.

OUACHITA MAP TURTLE

Graptemys ouachitensis Cagle 1953

Ouachita map turtle from Johnson County. Photo by Dean Hester.

IOWA STATUS. Protected; species of greatest conservation need.

HISTORICAL SUMMARY. The Ouachita map turtle has also been called the southern map turtle and was originally considered to be a subspecies of the false map turtle, named *Graptemys pseudogeographica ouachitensis* by Cagle (1953). Vogt (1978) first recognized *G. ouachitensis* as a species distinct from *G. pseudogeographica* and considered the species to have two subspecies—the more widespread Ouachita map turtle (*G. o. ouachitensis*) and the southern Sabine map turtle (*G. o. sabinensis*). Lindeman (2013) noted that the Sabine map turtle is allopatric—occurring in nonoverlapping geographic areas—nonintergrading, and identifiable and formally elevated it to a full species. As a result, only *G. ouachitensis* is currently recognized in Iowa.

Because the Ouachita map turtle was not described until the 1950s and thus early researchers did not recognize two distinct species, the early Iowa literature presented in the northern false map turtle historical summary may apply in part to the Ouachita map turtle. Vogt (1978) confirmed the identification of early Iowa map turtle specimens in the collections of the National Museum of Natural History–Smithsonian Institution and the University of Michigan Museum of Zoology. Based on this, the first confirmed specimen in the state appears to be in 1920 from the Mississippi River near Fairport in Muscatine County (National Museum of Natural History–Smithsonian Institution). However, another specimen in the National Museum of Natural History collection taken near Davenport in Scott County may have been collected earlier. No date is attached to the specimen, but based on the collection number it may been collected in the late 1800s or early 1900s.

DESCRIPTION. Ouachita map turtles have an olive or dark brown carapace with a distinct vertebral keel with black-knobbed projections at the back of the second, third, and fourth vertebral scutes and usually a dark blotch on the posterior edge of each scute. The carapace is covered with irregular yellow lines that somewhat resemble a map—*Graptemys* is from the Greek *graptos* meaning "inscribed" or "engraved" and *emys* meaning "turtle," possibly referring to the marks on the carapace that resemble a topographic map; *ouachitensis* is in reference to the Ouachita River where the type specimen—the specimen designated as the name-bearing "type" by the original author of the species to which the scientific name of that organism is formally attached—was collected. The yellow lines sometimes encircle the dark blotches. The posterior marginal scutes are serrated, the upper and lower surfaces of the marginal scutes have yellow circular markings at each seam, and the plastral bridge has longitudinal dark lines. The smooth hingeless plastron is cream to yellow and patterned with dark swirls that cover approximately half of the plastron in juveniles, though the pattern fades as the turtle ages and becomes indistinct in adults. The head, neck, and legs are olive to dark brown with yellow stripes. The neck stripes may or may not come into contact with the lower part of each eye. Hatchlings have a brighter carapace with a prominent vertebral keel and yellow or orange rings that fade with age and a dark pattern on the plastron.

The turtle is distinguished by three yellow markings on its head: a large square, rectangular, or oval mark located behind the eyes—sometimes forming a crescent—that is widest behind the eye and narrower on top of the head, a large spot under each eye, and another spot on the lower jaw directly below the eyes. Incubation temperature affects the formation of complete crescents, the size of the spots behind and beneath the eyes, and the number of lines entering the orbit of the eye, with cooler temperatures producing crescents, larger blotches, and fewer lines (Vogt 2018).

Plastron of a juvenile Ouachita map turtle from Linn County. Photo by Joshua Otten.

Ouachita map turtle from Louisa County showing the large mark behind the eye and the spots below the eye and on the lower jaw. Photo by Joshua Otten.

Ouachita map turtles exhibit pronounced sexual dimorphism, with males being smaller than females and having elongated foreclaws and longer, thicker tails with the vent located beyond the edge of the carapace; females are more than twice the size of males and have nonelongated foreclaws and shorter tails with the vent located beneath the carapace.

SIZE. Female Ouachita map turtles may range up to 9.4 inches (24 cm) straight-line carapace length; males generally range from 2.75 to 5.5 inches (7–14 cm) straight-line carapace length. The largest specimen in the Drake University Research Collection is a female from Muscatine County measuring 8.4 inches (21.4 cm) straight-line carapace length. The record straight-line carapace length reported for the species is 9.4 inches (24 cm).

SIMILAR SPECIES. Iowa's three *Graptemys* species may be found together and can be easily confused. The northern map turtle has a low vertebral keel lacking prominent spines, posterior marginal scutes that are less prominently serrated, and a small green to yellow triangular mark located just behind the eyes that is not connected to any of the neck or head stripes. The northern false map turtle, similar to the Ouachita map turtle, has a well-developed vertebral keel with prominent spines and more prominently serrated posterior marginal scutes, but it has a narrow yellow backward L–shaped mark behind its eyes and no spots on its lower jaw.

To add to the confusion, all three *Graptemys* species in Iowa are known to hybridize with each other. Studies quantifying molecular diversity in the species show that interspecies mating occurs (Freedberg and Myers 2012; Mitchell et al. 2016), and Vogt (1978) documented hybridization in the Mississippi River in Wisconsin. Species segregation may be assisted by different feeding niches and courtship behavior; nevertheless, turtles with characteristics intermediate among the three species have been found in Iowa.

DISTRIBUTION. The Ouachita map turtle is found primarily in the Mississippi River and its tributaries, extending from Minnesota and Wisconsin south to Louisiana, as far west as eastern Oklahoma and Kansas, and as far east as Ohio, West Virginia, Kentucky, and Tennessee. Many of the populations in tributaries east of the Mississippi River are not contiguous.

Ouachita map turtles are found in sites in eastern Iowa associated with the Mississippi River and some of its tributaries. Because the species was not described until the 1950s, it was not recognized by Reeve Bailey or other early herpetologists in the state. However, in the 1970s, Vogt (1978) examined museum specimens collected in Iowa and correctly identified those that are *G. ouachitensis*.

Based on this, the species had been documented in 4 counties by the mid-1940s; at present, records exist for 13 of Iowa's 99 counties. In the early 1980s, based on sampling conducted between 1970 and 1980, Jim Christiansen (1981) suggested that Ouachita map turtle populations in Iowa were unchanged. Joshua Otten (2020) greatly expanded the species' known distribution in the state with the first documentation in the Cedar River in Cedar, Linn, and Johnson Counties and the first documentation in the Skunk River in Lee County.

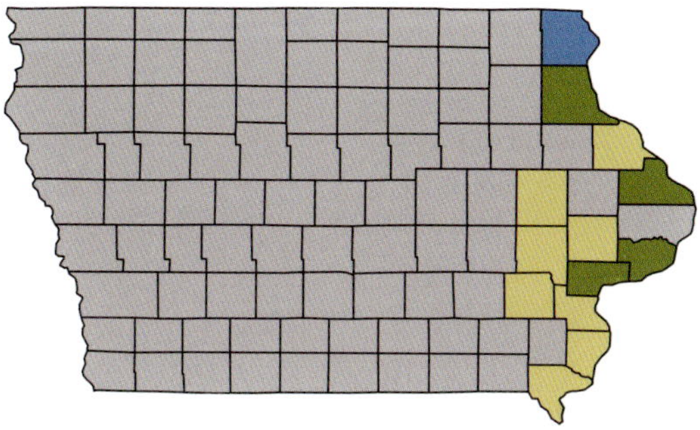

Ouachita map turtle distribution in Iowa.

Blue = pre-1960
Yellow = post-1960
Green = both pre- and post-1960

HABITAT. Ouachita map turtles are almost entirely a riverine species in Iowa and are found in medium to large rivers. They prefer areas with fast-moving water, abundant submerged vegetation, and a sand or silt bottom instead of the gravel, rock, or mud bottom preferred by Iowa's other map turtles. Numerous basking sites are also required because the turtle spends much of the day basking. Surveys in the Tennessee River drainage showed a significant positive relationship between the abundance of deadwood and basking densities of Ouachita map turtles (Lindeman 1999b). In the Sangamon River in Illinois, the turtles were found only in the larger downstream reaches of the river and were most often encountered in the lowest reach just prior to the confluence with the Illinois River (Bluett et al. 2013). Sixty-one percent collected as a by-product of fisheries monitoring in the upper and middle Mississippi River from 1992 to 1995 were captured in backwater habitats, though only a single individual was captured in the reach of the river adjacent to Iowa (Tucker, Gritters et al. 2008). In Kansas, the turtles were found to prefer deeper, wider sections of the river, above-average water temperatures, higher levels of dissolved oxygen, and less shade than sites preferred by either northern map turtles or northern false map turtles (Fuselier and Edds 1994).

BEHAVIOR. Similar to Iowa's other *Graptemys* species, Ouachita map turtles are shy, highly aquatic, and rarely found far from the water. If approached while

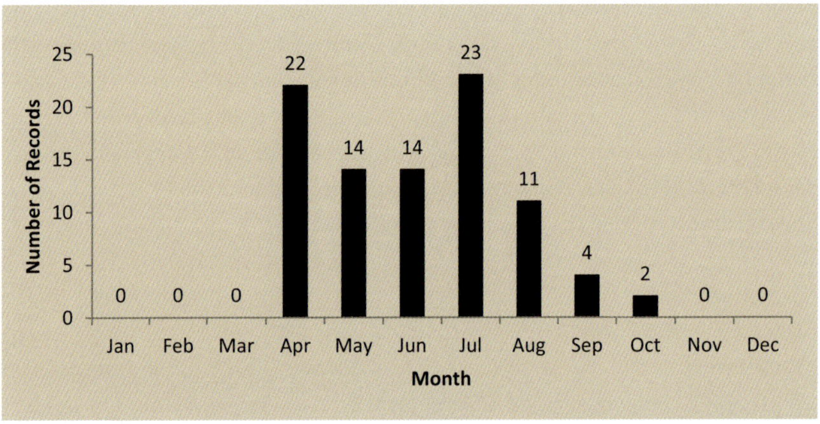

Collection/observation dates of 90 Ouachita map turtles in Iowa.

basking, they quickly slip or dive into the water. If handled, they rarely attempt to bite but may empty their bladders. Most will simply pull into their shells and wait for the threat to disappear.

The collection/observation dates of 90 Iowa Ouachita map turtles show that the species has been found from April to October in Iowa, similar to the active season in Wisconsin (Vogt 1980). The earliest Iowa record is from Dubuque County on April 13.

The behavior of the Ouachita map turtle is much like that of the closely related northern false map turtle, and like the northern false map turtle the Ouachita map turtle is almost entirely diurnal. When not foraging, the turtle is basking on muskrat lodges, on partially submerged logs, rocks, or stumps, or on sandbars not connected to the shoreline. It rarely basks on the shore, preferring basking sites surrounded by water. The turtles may bask alone or, more commonly, in groups, often with several turtles piled on top of each other. Vogt (1980) once observed 47 map turtles basking together on a log in Wisconsin. When in groups, they may not always get along; aggressive interactions occur with other Ouachita map turtles as well as with northern false map turtles and other species, with larger turtles frequently displacing smaller turtles from basking sites (Lindeman 1999a).

While basking, the turtles expose as much skin as possible to the sun by extending their necks, legs, and tails as far as they can, even spreading their toes to expose the webbing between. They are very skittish, and if one jumps into the water, often the others will quickly follow. Once the threat is gone, they cautiously return to the basking site. In addition to thermoregulation, basking can help reduce the amount of algae growing on the carapace, assist with digestion, and provide an opportunity for birds, such as common grackles, to assist in cleaning leeches from the turtles' bodies, with the turtle paying little attention to the pulling and pecking as the birds extract leeches from its skin (Vogt 1979). At night, the turtles sleep underwater on the bottom or on a submerged log.

The movement patterns of *G. ouachitensis* are not well known. In the Mississippi River in Wisconsin, Vogt (1980) recorded individual turtles traveling more than 2.5 miles (4 km) upstream and downstream during the active season and females traveling up to 1 mile (1.6 km) from river channels to nesting islands. Although there are many similarities in the behavior of the Ouachita map turtle and the northern false map turtle, there are some differences (Vogt 1980). Male Ouachita map turtles remain in river channels or sloughs during the summer, while both male and female northern false map turtles move into quiet backwaters to feed. Some female Ouachita map turtles also move into backwater areas, though many stay near the nesting islands to feed and bask, or they move into areas of aquatic vegetation immediately adjacent to the river channel.

During September, the turtles begin congregating at their overwintering locations. The latest Iowa record is one observed in Louisa County on October 1. Ouachita map turtles generally overwinter on the bottom of deep well-oxygenated pools where the water depth is sufficient to prevent freezing to the bottom.

Basking Ouachita map turtle from Johnson County. Photo by Mark H. Brown.

The turtles may overwinter singly or clustered in small or large groups and, like other map turtles, they may remain active throughout the winter. While they are submerged during the winter, through a combination of cloacal and bucco-pharyngeal respiration oxygen is extracted from the water and carbon dioxide from the blood.

THREATS. Raccoons, skunks, opossums, foxes, and river otters all eat the eggs of Ouachita map turtles and can be significant nest predators. A three-year study in Wisconsin documented a 90 percent nest depredation rate of Ouachita map turtles and northern map turtles, with the median time between egg deposition and nest depredation being 12.5 hours and 74 percent of the nests being depredated within 48 hours (Geller 2012). The larvae of a sarcophagid fly—a flesh fly—were observed attacking Wisconsin hatchlings in the nest (Vogt 1981c). Birds such as common grackles, American crows, and red-winged blackbirds prey on hatchlings making their way from the nest to the water; once in the water, the young are eaten by herons and large predatory fish like largemouth bass and northern pike (Vogt 2022c). Adults, with their highly aquatic lifestyle and wariness, are likely not commonly preyed upon, with nesting females being the exception. Raccoons were documented depredating adult females in Wisconsin as they laid their eggs (Geller 2020). Interestingly, although raccoons are generally active at night, some of the predation occurred during the day, with the raccoons foraging diurnally only during nest-construction periods.

Humans pose a significant threat to Ouachita map turtles through road mortality and habitat loss or alteration resulting from the modification of riverine habitats. In the 1990s, Terry (1993) studied the effects of stream modification on riverine turtles in Iowa, and although that work did not specifically address Ouachita map turtles, the closely related northern false map turtle was one species most vulnerable to stream modification. Ouachita map turtles are likely similarly affected by channelization, riprapping of streambanks, dam construction, and—because of their heavy reliance on woody debris as basking sites—any activity that removes deadwood from the channel.

Water pollution can affect the turtle as well. Studying the effects of the widely used pesticide atrazine on hatchling Ouachita map turtles and northern false map turtles from Muscatine County, Neuman-Lee and Janzen (2011) found that exposure to the lowest level of atrazine in the incubation substrate inhibited nest escape behavior and significantly reduced first-year survival in embryos. They concluded that atrazine-induced mortality in the wild was particularly harmful to long-lived species such as turtles, because if not enough hatchlings reach sexual maturity to replace those reproductive individuals that die, populations can decline, resulting in genetic bottlenecks, inbreeding depression, or both.

Commercial harvesting of map turtles, allowed in Iowa until 2006, may have posed a threat to some populations in the state. Although the harvest of map turtles was never as extensive as it was for snapping or softshell turtles, Iowa Department of Natural Resources records show that between 1987 and 2005—the last year that map turtle harvest was allowed—commercial trappers harvested an average of 325 pounds (147 kg) with a range of 0 to 2,054 pounds (0–932 kg) of map turtles per year in the state. Because harvesters were not required to distinguish among map turtle species, it is not possible to know exactly how many Ouachita map turtles were taken. However, the reported average does illustrate the magnitude of harvesting.

FOOD AND FEEDING. Ouachita map turtles are omnivores with a varied diet, including aquatic plants, snails, aquatic insects, crayfish, and to a lesser extent fish and other animal carrion. They often graze on the surface of the water, opportunistically feeding on whatever vegetation and insects they encounter, but they may also feed on the bottom by employing what has been called benthic bulldozing, during which large quantities of bottom material and insects are ingested (Moll and Moll 2000). They may also feed on terrestrial plants. Among members of a Mississippi River population, the most important plant food was turtlehead, a perennial terrestrial plant found in floodplain areas and various wetlands and along streams (Moll 1976). Although this turtle will eat mollusks, the narrow crushing surface of its jaws likely prohibits a diet heavy in mollusks.

Differences in diet may exist between males and females. In the Mississippi River in Wisconsin, plants such as pondweed, duckweed, eelgrass, and algae made up 40 percent of the diet of females (Vogt 1981b). Aquatic insects—mayfly, caddis fly, and damselfly larvae—made up another 44 percent followed by fish and carrion (5.2%) and mollusks (2.5%). Males were more carnivorous, eating the same insects as females along with beetles, flies, and other insect larvae, mollusks, and fish carrion but only traces of vegetation (Vogt 1981a). Moll (1976) found that males and females of approximately the same size had similar diets. In addition, he found that a change in the diet of a Mississippi River population occurred as the turtles grew. Animal food became less important as the turtles increased in size, and plant food was consumed more frequently and in a larger volume. Large adult females ate fewer but larger insects, such as adult beetles, wasps, and bees, rather than the smaller flying ants, beetles, and flies commonly eaten by smaller individuals. Moll (1976) also found that Ouachita map turtles living in the barren inorganic channel of the Mississippi River were more opportunistic, eating mostly insects that fell on the surface or were swept into the river during floods, whereas individuals living in the still waters of a lake adjacent to the Illinois River with a muddy organic bottom ate mostly midge larvae.

Ouachita map turtles are often sympatric—they occupy the same habitats—with Iowa's other map turtles. Each species reduces competition by using the environment in different ways that allow them to coexist—an example of niche segregation. The carnivorous northern map turtles specialize in mollusks, the omnivorous Ouachita map turtles are surface feeders, and northern false map turtles are generalist omnivores whose diet and feeding strategies overlap with those of the other two species (Vogt 1981a).

REPRODUCTION. Because Ouachita map turtle reproduction is similar to that of the northern false map turtle, the following comes primarily from the fine field studies of Vogt (1980) in Wisconsin except as noted. Males reach sexual maturity at around 4 to 6 years, while females mature later at around 8 years.

Females overwinter with a full set of vitellogenic follicles that are ovulated in the spring after emergence from overwintering, and females are able to store sperm over the winter. Mating takes place in the water in spring and fall when the turtles gather at their overwintering locations, with males likely recognizing females by head patterns or scent cues. During the courtship display, the male approaches the female from the front, arches both of his forearms near her head, and drums her ocular regions with the back of his long foreclaws. Each bout of drumming or vibration lasts around half a second, with the number of contacts per bout seeming to be species-specific—Ouachita map turtles average 5 and northern false map turtles average 10. Unlike the northern false map turtle, Ouachita map turtles do not bob their heads vertically as they drum with their claws. Once the female is motionless, the male moves into position and attempts to mount by making cloacal contact, hooking his tail around hers to maintain his balance, and bringing the vents together.

After they leave their overwintering sites, females travel to their nesting areas, where they spend several weeks basking as the eggs form shells. They exhibit strong fidelity to nest sites, returning to the same sites year after year. Nesting occurs on sandbars, islands, and beaches from mid-May to mid-July; the females dig flask-shaped nests 4 to 6.3 inches (10–16 cm) deep in soft sand or soil within 328 feet (100 m) of the water. Several test holes may be dug before the turtle selects a site. Along the Mississippi, lower Cedar, and lower Iowa Rivers, nests are often located on the sandy levees that border the rivers. Nesting occurs after dark or shortly after dawn but may also take place on humid overcast days, particularly after a rainstorm. Once the eggs are laid, the female fills in the hole with her back legs, after which she packs the sand by alternately raising each hind limb and then slapping the sand with it.

Females lay 2, possibly 3, clutches of 6 to 15 white elliptical leathery eggs per clutch per year; the eggs are 1.2 to 1.6 inches (3–4 cm) long. In Iowa, LeClere (2013) noted 11 Muscatine County clutches containing 6 to 11 eggs (average = 9) and a Muscatine County female that contained 9 fully developed and 3 infertile eggs. The eggs hatch from August to September after a 60- to 75-day incubation period. The nearly round hatchlings have prominent dorsal keels and average 1 to 1.4 inches (2.6–3.5 cm) straight-line carapace length. The 9 fertile eggs from the Muscatine County female noted by LeClere hatched on August 6 and 7. In Wisconsin, Vogt (1980) reported that 95 percent of the nests that survived the first 48 hours after laying hatched.

Well-developed embryos close to hatching and recent hatchlings have been shown to make sounds—principally clicks—within the nest (Geller and Casper 2019b). The precise function of these sounds is not known, but they may be some form of communication to help clutch mates synchronize hatching and nest emergence (Spencer and Janzen 2011). Hatchling emergence was mostly diurnal and synchronous—happening at the same time—in Wisconsin, occurring primarily in the morning soon after the soil temperatures began to rise from overnight lows (Geller et al. 2020b). Hatchling dispersal from the nest was nonrandom, with the hatchlings largely moving toward the nearest substantial vegetative cover (Geller et al. 2020a). Vogt (1980) found no evidence of overwintering in the nest in Wisconsin.

Like most turtles, Ouachita map turtles lack sex chromosomes, and the sex of hatchlings is determined by the temperature of the nest—temperature-dependent sex determination. In the lab, eggs incubated at 82.4° F (28° C) or below produced only males, and those incubated at 87.1° F (30.6° C) produced only females (Bull and Vogt 1979). In the wild, cool shaded nests produced only males, while those in the sun produced only females (Bull and Vogt 1979).

LONGEVITY. Both sexes have survived over 15 years in captivity (Ernst and Lovich 2009). Longevity in the wild is unknown.

NORTHERN FALSE MAP TURTLE

Graptemys pseudogeographica pseudogeographica (Gray) 1831

Northern false map turtle from Johnson County. Photo by Dean Hester.

IOWA STATUS. Protected; species of greatest conservation need.

HISTORICAL SUMMARY. The northern false map turtle has also been called the pseudo-geographic tortoise, Lesueur's map tortoise, Lesueur's terrapin, Lesueur's map turtle, midland sawback turtle, Mississippi map turtle, and Mississippi sawback turtle. The first confirmed museum specimens in the state appear to be in 1915 from the Mississippi River near Keokuk in Lee County and the Iowa River near Iowa City in Johnson County (National Museum of Natural History–Smithsonian Institution). However, Agassiz (1857) mentions receiving false map turtles from Burlington in Des Moines County, the *History of Fremont County, Iowa* (Anonymous 1881) listed Lesueur's map turtle as occurring in the county, and Somes (1911) stated that the species was found in Iowa but did not reference a specimen.

Two subspecies of the false map turtle, *Graptemys pseudogeographica,* are currently recognized, only one of which, the northern false map turtle, *G. p. pseudogeographica,* has a range that includes Iowa.

Nebraska, Iowa, Wisconsin, Illinois, Missouri, southwest Indiana, and northeast Arkansas. In Iowa, northern false map turtles are found associated with the Missouri River, the Mississippi River, and some of the larger eastern Iowa rivers. In the mid-1940s, the species had been documented in 7 counties; at present, records exist for 33 of Iowa's 99 counties. Joshua Otten (2020) expanded the known distribution in the state by locating the turtle in the Wapsipinicon River as far north as Buchanan County, the Cedar River north to Chickasaw County, the Des Moines River in Wapello County, and the Skunk River in Washington and Henry Counties.

Northern false map turtle distribution in Iowa.

Blue = pre-1960
Yellow = post-1960
Green = both pre- and post-1960

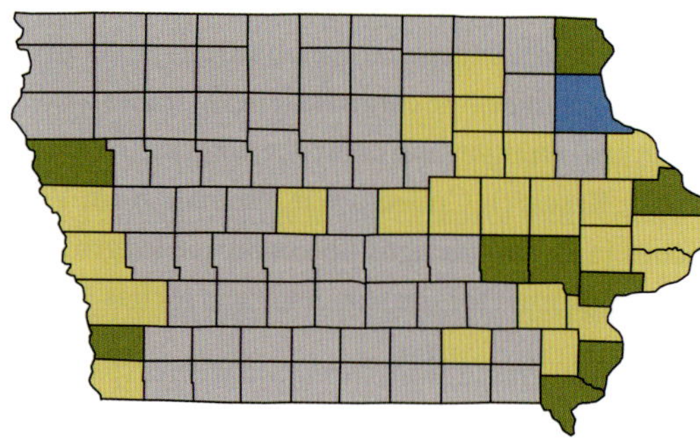

In the early 1980s, based on sampling conducted between 1970 and 1980, Jim Christiansen (1981) suggested that northern false map turtle populations in Iowa were unchanged. By the 1990s, however, Terry and Jim (1996) reported that northern false map turtles were among the turtle species most affected by stream modification in the state and that they had been eliminated or severely reduced in number in the most heavily modified rivers.

HABITAT. *G. p. pseudogeographica* is primarily a riverine species found in medium to large rivers in Iowa with adjacent open-canopy sloughs, backwaters, and oxbows; it generally prefers areas with slow-moving water, numerous basking sites, a muddy bottom, and some aquatic vegetation. Nevertheless, most northern false map turtles collected as a by-product of fisheries monitoring in the upper and middle Mississippi River from 1992 to 1995 were captured in main channel and side channel habitats, with relatively few collected in backwater or impoundment habitats (Tucker, Gritters et al. 2008). In the St. Croix River in Minnesota and Wisconsin, northern false map turtle abundance showed significant positive relationships with bank slopes and the percentage of muck in the substrate (DonnerWright et al. 1999). In the Missouri River in central Missouri, the turtle often

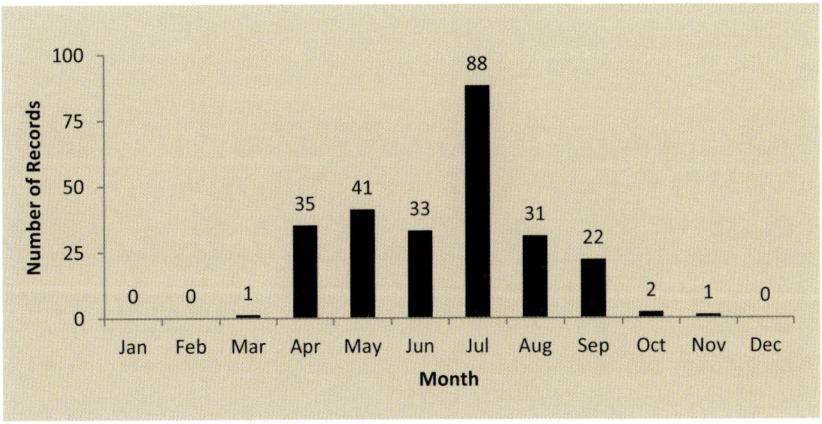

Collection/observation dates of 254 northern false map turtles in Iowa.

moves between the river and adjacent floodplain habitats, with females more often using the main river channel, males more often using the more permanent wetlands created by scouring by floodwaters, and hatchlings and juveniles using shallow and temporary wetlands (Bodie and Semlitsch 2000).

The turtles also inhabit lakes and reservoirs in some parts of their range, though Terry (1993) failed to detect them in Red Rock Reservoir in Marion County despite extensive trapping. He concluded that the unstable environment of the reservoir resulting from substantial fluctuations in water level reduced habitat diversity in the reservoir, which favored species more tolerant of disturbances. Likewise, Timken (1968), studying northern false map turtles in the Upper Missouri River, found the highest densities in the stable backwater habitats and the lowest densities in the ephemeral floodplains of large reservoirs.

BEHAVIOR. Like the other map turtles, northern false map turtles are shy, highly aquatic, and rarely found far from the water. If approached while basking, even from a distance, they will quickly slip or dive into the water. If handled, they rarely attempt to bite but may empty their bladders. Most will simply pull into their shells and wait for the threat to depart.

The collection/observation dates of 254 Iowa northern false map turtles show that the species has been found from March to November in Iowa, though they are primarily active from April through September. Vogt (1980) recorded their emergence from overwintering in April at water temperatures of 39.2 to 44.6° F (4–7° C) in the Mississippi River in Wisconsin. Carlander and Moorman (1951b) noted that northern false map turtles apparently do not hibernate as long as most other species and are usually one of the first turtles to appear in the spring and the last to disappear in the fall. The earliest Iowa record is one found in Linn County on March 21.

These turtles are almost entirely diurnal and, when not foraging, are basking, sometimes for many hours each day. They may bask singly or in large groups, often with other turtles such as northern map turtles, Ouachita map turtles, western painted turtles, and softshells. They can often be seen basking on muskrat lodges, partially submerged logs, rocks, or stumps, or sandbars not connected to the shoreline. They rarely bask on the shore, preferring sites surrounded by water. Basking turtles expose as much skin as they can to the sun by extending their necks, legs, and tails as far as possible and even spreading their toes to expose the webbing between them. When basking, they are very skittish, and if one jumps into the water the others often quickly follow. When the perceived threat is gone, they cautiously return to the basking site within about 10 minutes. In addition to thermoregulation, basking can help reduce the amount of algae growing on the carapace, assist with digestion, and give birds such as common grackles the opportunity to clean leeches from the turtles' bodies (Vogt 1979). The turtles pay little attention to the pulling and pecking as the birds extract leeches from their skin. At night, the turtles sleep underwater on the bottom or on a submerged log.

The movement patterns of northern false map turtles are not well known. Males move into quiet backwater areas to forage during the summer, while females seem to wait until they lay their first clutch of eggs before moving into backwaters to feed. In the Missouri River in central Missouri, no evidence was found to suggest that the turtles moved from one side of the 0.26-mile (425-m) wide river to the other (Bodie and Semlitsch 2000). In that same study, females used significantly more habitats and had greater average movement, average total movement, and average maximum movement than males. Throughout the year, the average maximum distance moved was 3.2 miles (5.2 km) by females and 2.4 miles (3.8 km) by males. In the Mississippi River in Wisconsin, Vogt (1980) felt that females potentially maintained regular home ranges because they were frequently caught in the same area each year. He also recorded the turtles moving more than 1.2 miles (2 km) upstream in less than a month and one female that moved approximately 5 miles (8 km) from her nesting site to her overwintering location in approximately 4 months.

As fall progresses, the turtles begin congregating at their overwintering locations, likely sometime in September or early October in Iowa. The latest Iowa

Basking northern false map turtle from Muscatine County showing the yellow backward L–shaped mark behind the eyes. Note the extended leg and the outspread toes. Photo by Elizabeth Owens/TNC.

record is one observed in Linn County on November 3. Interestingly, both the earliest and the latest records are from 2012, the hottest year on record as of 2022. During 2012, the average annual temperature was 4.5° F (2.4° C) above the long-term (1895–2020) average (Frankson et al. 2022). Typical of the Midwest, warming in Iowa has been concentrated in the fall and winter, which provides a possible explanation for the early and late dates for this species and illustrates how a changing climate has the potential to affect a species' activity period.

Northern false map turtles generally overwinter on the bottom of deep well-oxygenated pools where water depth is sufficient to prevent freezing to the bottom. In Wisconsin, overwintering in muskrat and beaver burrows and behind logs and rock piles was observed (Vogt 1981b). The turtles may overwinter singly or clustered in small or large, sometimes mixed-species groups. They may remain active during the winter and have been observed swimming under the ice; they may even bask on warm sunny winter days. When they are submerged in winter, they extract oxygen from the water and carbon dioxide from the blood through a combination of cloacal and buccopharyngeal respiration.

THREATS. Raccoons, skunks, opossums, foxes, and river otters all eat northern false map turtle eggs and can be significant nest predators. Common grackles, American crows, red-winged blackbirds, and other birds prey on the hatchlings as they make their way from the nest to the water; once in the water, the young are eaten by herons and by large predatory fish, such as largemouth bass and northern pike (Vogt 2022a). Adults, due to their highly aquatic lifestyle and wariness, are likely not commonly depredated, with the exception of nesting females who may be vulnerable while laying eggs. Vogt (1981b) stated that humans were the only known cause of adult mortality in Wisconsin; adult turtles drowned in gill nets, were captured on setlines, and shot.

Northern false map turtles are known to be hosts to a number of parasites, including leeches. Four coccidian parasites—single-celled endoparasites—were found in the Iowa northern false map turtles examined by Wacha and Christiansen (1976). Larvae of a sarcophagid fly—a flesh fly—were observed attacking Wisconsin hatchlings still in the nest (Vogt 1981c).

Humans pose a significant threat to northern false map turtles through habitat loss resulting from the modification of the riverine habitats in which they live. In the 1990s, Terry (1993) studied the effects of stream modification on riverine turtles in Iowa. The work showed that channelization, riprapping of streambanks, impounding of rivers, and other activities result in a reduction or loss of diversity of both the types of habitats and the turtle species in riverine systems. As the habitats become more homogeneous and less complex, rare species or those with specialized habitat requirements are lost. Northern false map turtles are very vulnerable to stream modification. Activities such as straightening, deepening, widening, and impounding rivers, along with removal of backwaters and oxbows, can all make rivers less suitable for these turtles.

Allowed in Iowa until 2006, commercial harvesting of map turtles may have harmed some populations in the state in the past. Although the map turtle harvest was never as intense as it was for snapping turtles or softshell turtles, based on Iowa Department of Natural Resources records between 1987 and 2005 (the last year map turtle harvest was allowed) commercial trappers harvested an average of 325 pounds (147 kg) with a range of 0 to 2,054 pounds (0–932 kg) of map turtles per year in Iowa. Commercial harvesters were not required to distinguish among map turtle species, and therefore it is not possible to know exactly how many northern false map turtles were taken. However, the average illustrates the magnitude of the map turtle catch.

Water pollution, road mortality, and winterkill are additional threats. In a study from Muscatine County of the effects of the widely used pesticide atrazine on hatchling northern false map turtles and Ouachita map turtles, Neuman-Lee and Janzen (2011) found inhibited nest escape behavior and significantly reduced first-year survival in embryos exposed to the lowest level of atrazine in incubation substrate. They concluded that atrazine-induced mortality in the wild was particularly problematic for long-lived species such as turtles; if not enough hatchlings reach sexual maturity to replace the reproductive turtles that die, populations can decline, resulting in genetic bottlenecks, inbreeding depression, or both. Winterkill is a concern in waterbodies where water levels have been reduced due to drought or other causes. Christiansen and Bickham (1989) documented winterkill of northern false map turtles in Louisa County.

FOOD AND FEEDING. Northern false map turtles are generalist omnivores that eat a variety of prey, including snails, aquatic insects, crayfish, and fish carrion. Aquatic plants, fruits, and seeds are also consumed. The turtles may occasionally feed on the surface but more commonly swim or crawl along the bottom, probing vegetation in search of prey. When prey is found, it is grabbed and swallowed whole, crushed with the jaws, or torn into pieces with the front feet.

In the Mississippi River in Wisconsin, females began feeding in the spring after laying their first clutch of eggs. Foods consumed (by volume) included vegetation (41.5%), insects (22.5%), mollusks (15.7%), fish (10%), and other unidentified material (9.5%), whereas males had a more carnivorous diet, consuming more aquatic insects (Vogt 1981a). Feeding for both sexes continued into September and ceased prior to overwintering. In Missouri, northern false map turtles were observed feeding on 13-year periodical cicadas that fell from the trees (Powell and Powell 2011). While on the surface, the turtles would raise their heads, apparently searching for the nearest cicada. When they spotted one, the turtles would quickly race to it before other turtles or fish could take it away.

Because northern false map turtles are often sympatric—occupying the same habitats—with the other map turtles in Iowa, competition among them is reduced because each species uses its food resources differently in a way that helps them to coexist—niche segregation. Northern map turtles are carnivores specializing in mollusks, Ouachita map turtles are omnivores specializing in surface feeding, and northern false map turtles are generalist omnivores that overlap the other two species both in diet and in feeding strategies (Vogt 1981a).

REPRODUCTION. Information regarding northern false map turtle reproduction comes primarily from the excellent field studies of Vogt (1980) in Wisconsin, which has been used for the following except as noted. Males reach sexual maturity at around 4 to 6 years; females mature later at around 8 years.

Females overwinter with a full set of vitellogenic follicles that are ovulated in the spring after the turtles emerge from overwintering, and they are able to store sperm over the winter. Mating takes place in the water in both spring and fall while the turtles are gathered at their overwintering sites, with males probably recognizing females by scent cues or head patterns. The courtship display involves the male approaching the female from the front, arching his forearms near her head, and drumming her ocular regions with the back of his long foreclaws. Each drumming or vibration bout lasts almost half a second, and the number of contacts per bout seems to be species-specific—with northern false map turtles averaging 10 and Ouachita map turtles averaging 5. When the female is motionless, the male moves into position and attempts to mount her by making cloacal contact, hooking his tail around hers to maintain his balance, and bringing their vents together.

After leaving their overwintering sites, females move to the nesting areas and spend several weeks basking as the eggs form shells. Females show strong nest site fidelity, returning to the same sites year after year. Nesting occurs from mid-May to mid-July, when females dig flask-shaped nests 4 to 6.3 inches (10–16 cm) deep in soft sand or soil within 328 feet (100 m) of the water. They may dig several test holes before selecting a site. Along the Mississippi, lower Cedar, and lower Iowa Rivers, nests are often located on the sandy levees that border the rivers. Nesting may occur throughout the day on overcast days but is concentrated in the morning on sunny days before the sand gets too hot. Rain may interrupt nesting, causing the female to return to the water. Once the eggs are laid, the female fills in the hole with her back legs, then packs it by alternately raising each hind limb and slapping the sand with it.

Vogt (1981b) considered large female northern false map turtles to be "egg factories," and he felt that if all the young turtles were to survive, "you could walk across the Mississippi River on the backs of turtles." Each female lays 2, possibly 3, clutches of 8 to 22 white elliptical leathery eggs per clutch per year; the eggs are 1.3 to 1.5 inches (3.2–3.7 cm) long. In Iowa, LeClere (2013) noted 13 Muscatine County clutches containing 7 to 13 eggs (average = 10).

The eggs hatch from August to September after a 60- to 75-day incubation period, and if the nests are not depredated, hatching success can be high. In Wisconsin, Vogt (1980) reported that 95 percent of the nests that survived the first 48 hours after laying hatched. Emergence from the nest occurs when the yolk sac is completely absorbed and may be asynchronous—that is, eggs may hatch over more than one day. Vogt (1980) found no evidence of overwintering in the nest in Wisconsin; however, in contrast, Christiansen and Gallaway (1984) reported

Hatchling northern false map turtle from Louisa County. Photo by Eric Osmundson.

both fall and spring emergence of northern false map turtle hatchlings in Louisa County. The nearly round hatchlings have a prominent dorsal keel and average 1 to 1.3 inches (2.6–3.3 cm) straight-line carapace length.

Like most turtles, northern false map turtles lack sex chromosomes; instead, the sex of hatchlings is determined by the temperature of the nest—temperature-dependent sex determination. Eggs incubated in the lab at 82.4° F (28° C) or below produced only males; those incubated at 87.1° F (30.6° C) produced only females (Bull and Vogt 1979). In the wild, cool shaded nests and those laid later in the season produced only males, while those in the sun produced only females (Bull and Vogt 1979).

LONGEVITY. A wild-caught adult survived in captivity an additional 32 years, 6 months, and 1 day. Longevity in the wild is unknown.

ORNATE BOX TURTLE

Terrapene ornata (Agassiz) 1857

Male ornate box turtle from Johnson County.

IOWA STATUS. Threatened; species of greatest conservation need.

HISTORICAL SUMMARY. The ornate box turtle has also been called the sand turtle, northern box turtle, painted box tortoise, and plains box turtle. Agassiz (1857) mentions receiving ornate box turtles from Iowa, though the first confirmed museum specimen in the state appears to be in 1884 near Davenport in Scott County (National Museum of Natural History–Smithsonian Institution). The *History of Fremont County, Iowa* (Anonymous 1881) listed the "northern box turtle" as very abundant in the county and the "common box turtle" (eastern box turtle, *Terrapene carolina carolina*), which does not occur in Iowa, as occurring as well. Osborn (1892) listed the ornate box turtle's range as "Iowa and west," Somes (1911) noted the turtle's presence in Iowa, and Brumfiel (1919) discussed its abundance in Johnson County. Blanchard (1923) reported an ornate box turtle found by a professor near Iowa Lakeside Laboratory in Dickinson County that was dissected and destroyed. He remarked that the report needed consideration, but the species was "undoubtedly very rare in the region." It is likely that the turtle was a released captive as no other records in the vicinity of Iowa Lakeside Lab exist. Similarly, Platt (1973) noted the eastern box turtle at Cone Marsh in Louisa County, but it is unclear if this was a released captive or a misidentified ornate box turtle. In either case, no other box turtle records in the vicinity of Cone Marsh exist.

Two subspecies of the ornate box turtle, *Terrapene ornata,* have recently been recognized, only one of which, the plains box turtle, *T. o. ornata,* has a range that includes Iowa. Martin et al. (2013) found no support for subspecies distinction between the plains box turtle and the desert box turtle, *T. o. luteola*. Likewise, the Turtle Taxonomy Working Group (2021) considered them the same species, as does Hughes et al. (in press).

DESCRIPTION. The ornate box turtle is a pretty turtle whose common name and specific and subspecific epithets come from the variegated colors on its carapace and plastron—*Terrapene* is from the Algonquin word for "turtle"; *ornata* is Latin meaning "embellished" or "fancy." The carapace is high-domed, round or oval, with a dark ground color and light to bold yellow lines on each scute that radiate downward and forward. Shell coloration and pattern vary in intensity. The top of the carapace is often flat or somewhat concave with the highest point above or slightly anterior to the plastral hinge. A middorsal keel is rarely present, and the carapace may flare at the margins in older turtles, though no serrations are present. The first marginal scute is irregularly oval or triangular, and there are usually 12 marginal scutes. The plastron is similarly marked with radiating yellow lines and has a strong single hinge, usually between the fifth and sixth marginal scutes, that creates movable anterior and posterior lobes permitting the turtle to tightly clamp its head, legs, and tail within its shell—hence the name box turtle—for protection. The skin is grayish brown to dark brown with yellow, orange, or red spots or blotches; the head and unnotched upper jaw are yellow or green. The hind feet have 4—rarely 3—toes. Hatchlings have light brown shells with some yellow spotting on the carapace and tan skin.

Males tend to be more dorsoventrally compressed—flattened from the upper as well as the lower surface—have thicker first hind toes and claws that are turned inward for clasping females during copulation, slightly concave plastrons, and longer tails with the vent located beyond the edge of the carapace, and they can have a green head and upper jaw. During breeding, males develop bright red irises and red coloration on their forelimbs, though the coloration fades later in the year. Females have yellowish brown irises, first hind toes that do not turn inward, and flat or convex plastrons.

Plastron of an ornate box turtle from Linn County.

SIZE. Ornate box turtles generally range from 4 to 5 inches (10–12.7 cm) straight-line carapace length. The largest specimen in the Drake University Research Collection is from Louisa County measuring 4.7 inches (12 cm) straight-line carapace length. Several studies indicate females are slightly larger than males; however, data suggest that this is not the case in Iowa. Measurements from 1993 to 2015 by Neil and his coworkers in Johnson County indicate that the largest male was 4.9 inches (12.5 cm) straight-line carapace length, while the largest female measured 4.8 inches (12.2 cm). Average straight-line carapace length for adults with more than 9 growth rings was 4.3 inches (10.9 cm; n = 337) for males and 4.2 inches (10.7 cm; n = 370) for females. In Louisa County, measurements by Jim Christiansen, Neil, and their coworkers from 1995 to 2007 show that some individuals were notably larger than those in Johnson County. The largest male had a straight-line carapace length of 5.2 inches (13.2 cm), while the largest female measured 5.4 inches (13.7 cm). The average straight-line carapace length for adults with more than 9 growth rings was 4.5 inches (11.5 cm; n = 206) for males and 4.3 inches (11 cm; n = 289) for females. The record straight-line carapace length reported for the species is 6.1 inches (15.4 cm).

SIMILAR SPECIES. Blanding's turtles have a single plastral hinge but have a dark brown to black carapace with white to yellow speckling or irregular radiating lines, a yellow plastron with dark blotches—though it may be almost totally black on some individuals—and a bright yellow neck and chin. Yellow mud turtles have 2 plastral hinges, are less colorful, and have 11 marginal scutes. Two nonnative box turtles—the eastern box turtle and the three-toed box turtle (*T. c. triunguis*)—are commonly released in Iowa and may be confused with ornate box turtles. Eastern and three-toed box turtles have middorsal keels, nearly rectangular first marginal scutes, and plastrons that lack bold radiating yellow lines. Three-toed box turtles tend to have 3 toes on each rear foot; ornate box turtles usually have 4.

DISTRIBUTION. Ornate box turtles extend from isolated populations in southwest Wisconsin, Illinois, Indiana, and Iowa south through Missouri, Arkansas, and Louisiana and west to southern South Dakota, Nebraska, Kansas, eastern Colorado, Oklahoma, Texas, and eastern New Mexico. While populations were once large and continuous throughout the Great Plains, most are now separated by human barriers (Richtsmeier 2005; Richtsmeier et al. 2008). Winter temperatures that cause the ground to freeze deep enough to kill hibernating turtles seem to be a factor limiting the northern distribution (Legler 1960).

T. ornata probably spread into Iowa during the xerothermic period—a warm dry postglacial interval approximately 7,000 to 5,000 years before the present—which encouraged reptile migration as grasslands replaced woody vegetation over central North America. The distribution of Iowa's more arid-adapted reptiles, like the ornate box turtle, can be explained in part by former connections to populations in xeric grasslands to the west and southwest of Iowa.

In the mid-1940s, the species had been documented in 9 counties; at present, records exist for 37 of Iowa's 99 counties, 5 of which have only pre-1950 records. Prior to intensive agriculture, ornate box turtles were likely more widely distributed across the state. Along with populations in the Loess Hills of western Iowa, many sandy prairie areas throughout central and eastern Iowa were inhabited, and remnant populations in southern Iowa indicate a once-wider distribution. In the early 1980s, Jim (1981) concluded that ornate box turtles were rare in all but a few places in Iowa. The Loess Hills probably provided the most contiguous ideal habitat for the species, though in the mid-1980s Jim and Catherine Mabry (1985) did not encounter the turtle during their surveys of the Loess Hills. The turtle is now rare in the Loess Hills, though individuals are still occasionally found. The largest remaining populations occur in Johnson, Louisa, and Muscatine Counties.

HABITAT. Ornate box turtles are primarily a prairie species with a preference for sand prairies, but at times they also inhabit pastures, cultivated land, wet woods, and open woodlands. The turtles are restricted to areas with soils that allow for

Ornate box turtle distribution in Iowa.
Blue = pre-1960
Yellow = post-1960
Green = both pre- and post-1960

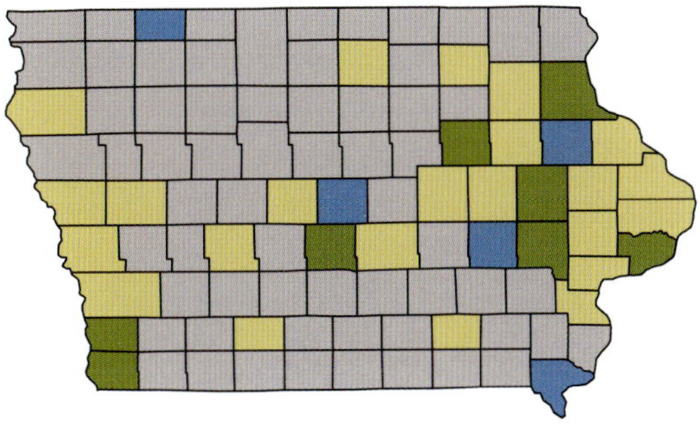

easy burrowing. Loose sand is essential so they can bury themselves during the winter, for thermoregulation the rest of the year, and for digging nests. In Kansas, Fitch (1958) reported the order of preference for habitats as grazed pastures, woodlands, open fields with undisturbed prairie vegetation—smooth brome and weeds were less preferred—and fallow fields with a rank growth of weeds being the least preferred. Legler (1960) also reported the ornate box turtle's preference for pastures, especially where domestic cattle provided an abundant supply of dung that attracts dung insects—an important food. During a Linn County study, Terry and his colleague Stacey Parks found a small isolated population of ornate box turtles using overgrazed pastures, farmyards, and croplands. Several individuals used a junkyard that, while overgrazed, had sandy soils and abundant cow dung, and the old equipment provided ample cover.

During long-term studies in Johnson County, Neil and his coworkers found that all age groups use mainly open prairie and shrub borders of prairie from May to September (Bernstein, Fendrich et al. 2023). With the exception of first-year individuals, the turtles disperse from the open prairie to shrubs, forests, wetlands, or agricultural fields during June and July and then return to the prairie and shrub borders by August and September prior to overwintering.

BEHAVIOR. Ornate box turtles are generally calm if handled. Some will completely withdraw into their shells while others may withdraw only partially or not withdraw at all. If handled, most will usually only thrash their feet to escape, though some may bite. When they do bite, they clamp onto the skin with their powerful jaws and do not readily release. Neil was once careless with a three-year-old that he believed to be relatively harmless, but he was taken to the ground when it chomped the fleshy part of his finger and would not let go until forcefully pulled off—not without damage to his finger. Some may also empty their bladders when frightened, which can stress the turtle during dry conditions. Laboratory

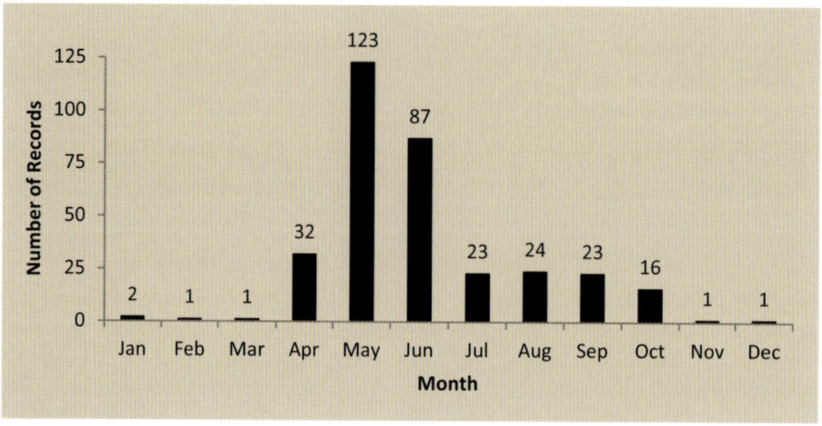

Collection/observation dates of 334 ornate box turtles in Iowa.

studies have demonstrated the turtles' learning ability (Rosenbaum 1968; Harless and Lambiotte 1971), and Reed et al. (2023) found that three behavioral traits—boldness, exploration, and activity—did not differ between sexes or sizes of turtles within a population but were different between populations.

The collection/observation dates of 334 Iowa ornate box turtles show the species has been found in the state every month of the year, though the primary active season is April through October. The earliest record is from Benton County on January 10; the latest record is December 3 in Cedar County. The turtle is most active from April to June (74% of records), with 37 percent of the records from May. In Johnson County, some individuals begin emerging from overwintering as early as April 1, with most emerged by mid-April (Bernstein and Black 2005). Emergence was thought to be triggered by soil temperature (Grobman 1990), but Neil and Robert Black (2005) found that soil temperature alone was not enough to predict emergence, which was corroborated by Edmonds et al. (2023), who found that precipitation was also a factor.

Except for females during nesting, the turtles are mainly active during the morning and late afternoon when they bask and feed. They spend much of the day and night partially burrowed into shallow semicircular holes called forms, under vegetation at the base of shrubs and bunchgrasses, in mammal burrows, or under logs. Most individuals in Johnson County were found burrowed under vegetation and shielded from direct sunlight by grasses, shrubs, or trees (Bernstein, Fendrich et al. 2023), thereby preventing overheating. While they are largely terrestrial, Legler (1960) reported individuals submerged in muddy water during the hottest

Male ornate box turtle in a form in Johnson County.

part of the day, and Neil has seen the turtle "swim" by walking along the bottom of shallow water. Ornate box turtles are also good climbers and can navigate over fallen vegetation, traverse steep banks, and even climb fences in captivity.

Tucker et al. (2015) found males to be generally more active than females, reduced activity of both sexes during midday, and an increase in activity during rain events. They concluded that thermal constraints determine turtle activity and that warm dry years inhibit movement. They also found that relative humidity, reproductive status, and time of day and year influence activity.

Neil and his coworkers (Bernstein et al. 2007) found that male annual home ranges in Johnson County were larger than those of females and that both sexes were most active in May and June during the breeding and nesting seasons. Breeding and nesting occurred near the overwintering area. After breeding, some turtles dispersed to a summer range where they remained sedentary until they returned to their overwintering location, thus displaying a "dumbbell home range" where one area is used for overwintering and breeding and another is used the rest of the year. There is much individual variation in adult annual home ranges. In Johnson County, total home range varied between 2.2 and 11.6 acres (0.9–4.7 ha) with the core area ranging from 0.001 to 1 acre (less than 0.001–0.4 ha). Individual turtles displayed a high degree of home range fidelity and, for the most part, were found in the same area at the same time of year during different years and nested and overwintered in relatively the same locations each year.

Information about the different breeding and summer ranges was not known at the start of the Johnson County study, so Neil was surprised one day when a student lost the signal of a transmitter placed on a turtle the day before. With careful listening, they were able to get a faint radio signal and discovered that the turtle had moved about 0.6 mile (1 km) in less than 48 hours over shrubs and logs and through a drainage ditch. The turtle remained on its summer range, barely moving, until September when it returned to the same spot where it was first encountered, again traveling the distance in fewer than 48 hours. In contrast, another individual barely moved 82 feet (25 m) the entire year.

As fall approaches, the turtles return to their overwintering spots and travel less as they prepare for winter. Unlike Iowa's aquatic turtles, ornate box turtles overwinter on land, where they burrow into sandy or loamy soils in areas with short sparse vegetation. Neil and Robert Black (2005) found that fall burrowing began by the third week of September in Johnson County, with most turtles belowground by the first or second week of October, resulting in an overwintering period ranging from 176 to 201 days. During the winter, most presumably avoid freezing temperatures by burrowing deeper as the frost line descends, and Neil and Black (2005) documented a maximum overwintering depth of 5.5 feet (1.67 m) with an average depth of 2.5 feet (0.75 m). They also found that 2 shallow-buried individuals survived over 3 weeks of freezing temperatures one winter. Ornate box turtles generally overwinter alone. However, in 2004 Neil encountered 4 newly emerged individuals—each with 6 growth rings—clustered together on April 22, and because of their uniform age he speculated that they were nest mates that, potentially, had overwintered together. He has also encountered multiple turtles—usually no more than 3—overwintering together in the same burrow.

THREATS. Ornate box turtle eggs are eaten by a variety of predators, including raccoons, striped skunks, opossums, plains hog-nosed snakes, and bullsnakes. Most nest predation happens within 24 to 48 hours of laying, if not the same night. One morning, Neil found a female lying over a shallow depression whose legs had been chewed off as she was laying eggs the night before. In the years since 1994, when Sandy Rhodes began monitoring the Johnson County population, only rarely have hatchlings been encountered that were not from nests protected by cages. In Wisconsin, most nest predation happened closer to shrub edges (Temple

1987); however, Neil and his fellow researchers (2015) were not able to replicate that result using artificial nests in Iowa. What was found, however, was that mesopredators were cued by soil disturbance from digging the nests rather than by the eggs themselves. Geller (2015) came to the same conclusion, suspecting that the cue was an odor from a soil fungus released after digging.

Adults are less prone to depredation due to their strong shells and hinged plastrons; however, they are not immune. The weaker-shelled juveniles are vulnerable to crows, red-tailed hawks, raccoons, coyotes, domestic dogs, and other predators. Hatchlings are preyed upon by small predators, such as shrews, as well as snakes and large birds. In Johnson County, a hatchling was found depredated with what appeared to be tooth marks from a short-tailed shrew, and another hatchling was missing a foot. Predators like shrews may just gnaw the feet off vulnerable hatchlings; adults are sometimes found with healed foot amputations.

Jim and his coworkers (2004) documented reoccurring facial lesions on ornate box turtles in Louisa County, though the turtles had no apparent harm; in Kansas, Legler (1960) noted various ectoparasites—parasites that live on the outside of another animal—again without apparent harm. However, Iverson (2023) documented that botfly infestations can be a significant cause of mortality in some years; see Hughes et al. (in press) for a more complete discussion of disease studies. As early as the 1930s, Scott (1938) documented road mortality of ornate box turtles in Clayton and Hamilton Counties; this is still a concern in some locations, as is mortality due to farm machinery. Genetic analysis indicates that ornate box turtles and eastern box turtles can hybridize (Martin et al. 2020), illustrating the concern over releasing nonnative eastern box turtles into Iowa, which may promote hybridization and introduce disease.

Ornate box turtles are sand prairie specialists, and sandy soils are critical for burrowing and for nest building. Habitat loss has undoubtedly affected their distribution in Iowa. Once relatively widespread, this species has become restricted to the few remaining areas of suitable habitat. As sandy areas are cultivated and urbanized, ornate box turtles and the other species that depend on these special habitats lose their abilities to survive and reproduce. In the late 1990s, Jim (1998) concluded that the loss of open prairie and its more modern substitute, permanent pasture, had contributed to declines of several prairie-adapted reptiles. Urban sprawl poses a threat as well. From 1998 to 2003, an ornate box turtle population was relocated from a site on the south side of Iowa City in Johnson County that was under threat of urban development to a location in Polk County (Hill et al. 2009).

Improper management of sand prairies can be harmful as well. In addition to outright loss of habitat, woody invasion of sand prairies is a concern, either through natural succession or by well-intentioned land managers planting conifer trees or dense tall grasses on sparsely vegetated sand prairies to prevent erosion. Fire can be a useful tool for managing woody species, but burning prairies at the wrong time of the year or under the wrong conditions can have dire consequences for the reptiles living there. Paul Frese (2003) documented high reptile mortality after a large undisturbed prairie in Missouri was burned. Out of 71 individual reptiles and 8 species, he found that 43 (60.5%) had died as a result of the fire, including 4 ornate box turtles and 20 three-toed box turtles. This example clearly illustrates the need to conduct prescribed fire outside of the reptiles' active season—from November to March—to avoid mortality, though as the climate warms the time window for safe burning will likely get shorter.

Because of their attractiveness and temperament, ornate box turtles are popular in the pet trade, and collection as pets has affected local populations in some areas. Although prohibited in Iowa, illegal collection still occurs. A Johnson County population was significantly reduced by a poacher removing turtles for the pet trade. Studies have shown that few of these turtles survive long in captivity; of those that do, many are released far from where they were captured. Because box turtles have an excellent homing ability, once released they will attempt to make the perilous and often fatal trip to "return home."

FOOD AND FEEDING. Ornate box turtles are omnivores that find food by both sight and smell. Live prey may be chased; larger food items are torn into smaller pieces with the front feet or jaws before being eaten. The turtles eat a variety of larval and adult insects, earthworms, snails, and other invertebrates, fish and amphibians, carrion, and a variety of plants and berries. When strawberries, blackberries, or mulberries are in fruit, individuals can often be found with purple-stained mouths. We have both radio-tracked ornate box turtles who spent much of their time under mulberry trees, and one readily nibbled on mulberries offered by hand during transmitter attachment. They are more carnivorous than the eastern box turtle, with insects such as dung beetles, caterpillars, and grasshoppers accounting for approximately 90 percent of their diet (Legler 1960).

REPRODUCTION. Maturation in ornate box turtles is more closely related to size than to age. In general, males grow faster and become sexually mature when smaller and younger, maturing in 7 to 9 years at straight-line plastron lengths of 4 to 4.3 inches (10–11 cm), whereas females mature in 10 to 11 years at straight-line plastron lengths of 4.3 to 5.1 inches (11–13 cm) (Legler 1960; Bernstein et al. 2018). The ovarian cycle begins in July shortly after egg laying and continues until the following spring, when the follicles enlarge rapidly before ovulation in late April or May.

Ornate box turtle from Louisa County eating a toad. Photo by Ryan D. Rasmussen.

Mating usually occurs in the spring—late April through May in Iowa—though it may also occur in the fall. Once Neil placed a male in an aquarium with a female in May, and they were copulating within 5 minutes. The great herpetologist Archie Carr (1952) wrote, "There is no sense in suppressing the comment that a pair of copulating box turtles is ludicrous to behold." During courtship, the male approaches the female, often from the rear while probing, biting, or pushing her shell, and eventually launches himself onto her back, somewhat struggling to grasp her. He grasps the rear underside of the female's carapace with his claws, or he may use his larger inner hind claws to grip her legs or skin. The female wraps her legs around the outer legs of the male, pressing his legs inward while she rocks backward, thus securing his position. During copulation, the male leans backward—almost upside down. Copulation can last up to 2 hours before the female relaxes her plastron, releasing the male. Both sexes may mate several times during a season with multiple partners. Females can store sperm from one season to the next.

Nesting occurs primarily in the evening or at night from late May to late June, peaking during the first 2 weeks of June in Iowa. All Johnson County nesting observations have been at night. Females travel to their nesting areas several days ahead of time and dig several test holes before choosing a site. Several studies indicate that they construct nests and lay eggs during or after thunderstorms.

Mating ornate box turtles in Linn County.

Once while conducting an ornate box turtle survey in Tama County in early June, Terry jumped in the car after seeing that thunderstorms were predicted. Arriving at dusk just ahead of the storms, he quickly found a female starting to dig a nest.

Nest sites typically are bare well-drained sloping areas. Johnson County nest sites are on relatively open sand prairies rather than in tall grasses or under shrubs. When digging a nest, the female scoops sand from under her body by using her rear legs alternately to excavate a flask-shaped nest chamber with a narrow neck approximately 2.2 inches (5.5 cm) deep, 3.1 inches (8 cm) wide at the bottom, and 2.4 inches (6 cm) wide at the opening (Legler 1960). The female may urinate in the nest after depositing eggs as well as urinate on the nest after covering it with sand. While it has been suggested that ornate box turtles can lay 2 clutches in a year, studies in Johnson County have not observed that, and data suggest that females lay, at most, every other year. Only a fraction of females lay eggs each year (Doroff and Keith 1990; Redder et al. 2006); we do not know of any instance in Iowa where females have laid more than one clutch per year. Ornate box turtles have temperature-dependent sex determination—with lower incubation temperatures producing males and higher temperatures producing females (Vogt and Bull 1982).

Across the range of the species, clutch size varies from 1 to 8 eggs, with 4 to 5 being the usual number. The eggs are white, ellipsoidal, 1.4 inches (3.5 cm) long,

Hatchling ornate box turtle from Louisa County. Photo by Ryan D. Rasmussen.

smooth to finely granulated, and brittle. In the Midwest, mean clutch sizes of 4.7 eggs (Kansas, Legler 1960), 2.6 eggs (Nebraska, Converse et al. 2002), 3.5 eggs (Wisconsin, Doroff and Keith 1990), and 4 eggs (Wisconsin, Hay and Christoffel 2022) have been reported. In Iowa, Neil and his coworkers calculated a mean clutch size of 4.4 eggs (range 2 to 6) in Johnson County. Two Louisa County females in the Drake University Research Collection, collected on May 30 and May 31, contained 5 and 4 oviductal eggs, respectively.

Eggs hatch in August and September, but the hatchlings stay in the nest until the following spring. The hatchlings, 1.1 to 1.3 inches (2.8–3.2 cm) straight-line carapace length, are highly tolerant of freezing and can survive temperatures as low as 26.6° F (–3° C) for up to a week (Costanzo et al. 1995).

LONGEVITY. Wild ornate box turtles can live over 50 years (Lewis and Iverson 2018). Jim and his coworkers (2004) documented a Louisa County female that was at least 37 years old. Based on recaptures in 2022 of turtles marked by Sandy Rhodes from 1993 to 1996 in Johnson County, some individuals have lived more than 40 years (Bernstein, McCollum et al. 2023). It is probable that some Johnson County turtles are over 50 years old, and a few individuals marked by Jim in Louisa and Muscatine Counties in the 1970s and 1980s have been recaptured in the last 10 years (Joshua Otten, personal communication, 2023).

RED-EARED SLIDER

Trachemys scripta elegans (Wied-Neuwied) 1838

Red-eared slider from Wapello County. Photo by Ryan D. Rasmussen.

IOWA STATUS. Protected.

HISTORICAL SUMMARY. The red-eared slider has also been called Troost's turtle, Troost's terrapin, Cumberland terrapin, pond slider, and red-eared turtle. The first confirmed specimen in the state appears to be in 1916 from Keokuk in Lee County (National Museum of Natural History–Smithsonian Institution). Five specimens were collected near Burlington in Des Moines County in the 1920s (Museum of Comparative Zoology–Harvard University). Carr (1938) examined a specimen from Burlington, and Conant (1938) indicated all of southern Iowa as being within the range of the species.

Three subspecies of the pond slider, *Trachemys scripta*, are currently recognized, only one of which, the red-eared slider, *T. s. elegans*, has a range that includes Iowa.

DESCRIPTION. The red-eared slider is a medium-sized turtle whose appearance gives rise to both its common and scientific names. *Trachemys* is derived from the Greek words *trachys* meaning "rough"—in reference to the rough surface of the carapace—

and *emys* meaning "turtle"; *scripta* is Latin meaning "written" or "marked"—in reference to the shell pattern; *elegans* is Latin meaning "fine" or "elegant"—in reference to the red stripes behind the eyes. The olive to dark brown carapace is oval and weakly keeled with a wide yellow stripe on each pleural scute, numerous smaller black and yellow lines, and a weakly serrated back edge. A yellow line is present in the center of the upper surface of each marginal scute, and the underside of each marginal scute has a dark blotch surrounded by a light band. The smooth hingeless plastron is yellow with a large dark blotch near the center of every scute, and the plastral bridge is patterned with dark blotches or bars. The skin is green to olive-brown, and the neck, legs, and tail are all striped with yellow lines. The most prominent feature on the head is a wide red stripe behind the eyes. In addition, there are narrow chin stripes and a wide yellow stripe extending downward and backward from the eyes. Hatchlings have a pattern similar to that of adults—a red stripe behind the eyes and large circular blotches on the plastron—but have a distinctly green carapace with a vertebral keel.

Adult males and the plastrons of a few females become darker—melanistic—as they age. The spreading of the excess dark pigment covers the plastral pattern, hides most of the yellow stripes on the carapace and skin, and even obscures the

Plastron of a red-eared slider from Linn County.

red stripes behind the eyes. The difference in appearance between melanistic and nonmelanistic males is so striking that at one time they were thought to be different species. Interestingly, though males are adults before the change begins, the timing of the onset of melanism in males is dependent on population-specific female body size at maturity rather than male age or size (Ernst and Lovich 2009).

Males have flatter carapaces, elongated curved foreclaws, and longer and thicker tails with the vent located beyond the edge of the carapace, whereas females have higher, more domed carapaces, short straight foreclaws, and shorter tails with the vent located beneath the carapace.

SIZE. Red-eared sliders generally range from 5 to 7.9 inches (12.7–20 cm) straight-line carapace length. The largest specimen in the Drake University Research Collection is a female from Louisa County measuring 9.6 inches (24.3 cm) straight-line carapace length. The maximum straight-line carapace length reported for the species is 11.9 inches (30.2 cm).

SIMILAR SPECIES. Among Iowa's other turtles with yellow stripes on the neck and head, none has a broad red stripe behind the eyes. The three map turtle species have yellow spots on the head and a distinct vertebral keel, and the back edge of the carapace is strongly serrated. Western painted turtles have a plastron that is yellow-orange, bright orange, or red and patterned with variably sized black to reddish brown markings, a plastral bridge that is unmarked, and no serration along the back edge of the carapace. With the exception of the map turtles and the western painted turtle, all other Iowa turtle species lack yellow stripes on the neck, head, and legs.

DISTRIBUTION. The red-eared slider's native range includes the Mississippi River valley from southeast Iowa and Illinois south to the Gulf of Mexico, as far east as West Virginia, and as far west as eastern New Mexico. However, red-eared sliders are the most popular pet turtle in the United States and are very popular elsewhere in the world, resulting in thousands, likely tens of thousands, of former pets being released into wetlands and streams around the world—often with dire consequences for native turtles. A recent California study implicated introduced red-eared sliders as a vector of *Emydomyces testavorans*—a fungus capable of infecting the shells of freshwater turtles (Lambert et al. 2021). The species has been introduced in all 50 states, all continents except Antarctica, and some oceanic islands, making it the world's most widespread freshwater turtle and ranking it among the world's 100 worst invasive alien species according to the International Union for Conservation of Nature.

The red-eared slider's native Iowa range was limited to very southeast Iowa. In the mid-1940s, the turtle had been documented in just 3 counties. Carlander and Moorman (1951b) stated that it was found only in the Mississippi River in Des Moines and Lee Counties. In the early 1980s, Jim Christiansen (1981) considered it to be threatened due to its limited distribution, but by the late 1990s (1998) he had changed his assessment, noting that the turtle had recently been found in abundance in all major southeast Iowa rivers and was now considered common.

Red-eared slider distribution in Iowa.
Yellow = post-1960
Green = both pre- and post-1960

At present, this turtle has a scattered statewide distribution due to introductions across Iowa, with records in 21 counties, though it likely occurs elsewhere as well. The largest populations are still in southeast Iowa, where they are common in the Mississippi River and its backwaters as far north as Davenport in Scott County, and they are also found in the lower Des Moines, Iowa, Cedar, Skunk, and Wapsipinicon Rivers.

HABITAT. Red-eared sliders are found in quiet portions of rivers, lakes, farm ponds, marshes, sloughs and backwaters, drainage ditches, urban stormwater detention basins, and backyard ponds. River floodplains are used heavily; aquatic habitats with mud bottoms, abundant aquatic vegetation, and plentiful basking sites are preferred. In southeast Iowa, the turtles are most often found in large ponds or small lakes, backwater sloughs and oxbows, gravel pits, or inlets of large rivers. Braun and Phelps (2016) reported that red-eared sliders were captured most often in tributaries and closed side channels of the Mississippi River. In agreement, Bodie and Semlitsch (2000) noted that red-eared sliders in the Missouri River were captured only in off-channel habitats, though they also used the main channel. They spent much of the year in the river, especially during the cooler months, and they used more diverse habitats in the warmer months (Bodie and Semlitsch 2000).

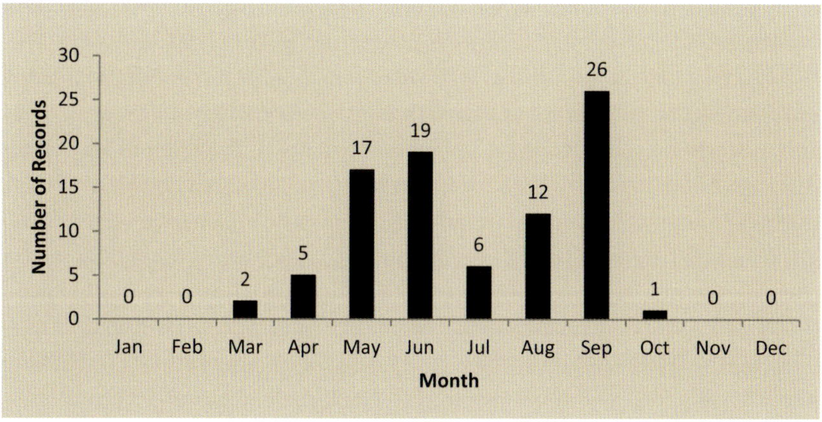

Collection/observation dates of 86 red-eared sliders in Iowa.

Habitat use differs between sexes and age groups (Bodie and Semlitsch 2000). Males are more sedentary, use half as many habitats, and move half as far from the river onto the floodplain as do females, especially from May through July. Juveniles are often found in aquatic habitats farther from the river than adults, possibly reflecting the close proximity of nesting areas to shallow wetland habitats with ample food and a lack of aquatic predators.

BEHAVIOR. Red-eared sliders are skittish and, if approached while basking, will quickly slide into the water and swim away—the name slider comes from this behavior of sliding into the water when startled. If handled, they are generally passive, but some may bite. They may also empty their bladders and kick and scratch as they attempt to "swim" away. Aggression against other red-eared sliders occurs at basking sites, sometimes forcing individuals off of basking platforms, and melanistic males are more aggressive than nonmelanistic males (Lovich et al. 1990).

The collection/observation dates of 86 Iowa red-eared sliders show that the species has been found from March to October in Iowa, though most activity is from April through September. The earliest Iowa record is a collection by Reeve Bailey on March 22 near Montrose in Lee County. In Illinois, red-eared sliders did not become active in the spring until the water temperature reached 50° F (10° C) or higher (Cagle 1950). In Iowa, they are actively feeding by May and continue foraging into early September, at which time feeding begins to slow down as they prepare for winter. Jim noted that he was never able to catch them in baited traps in either September or October, though he frequently had his herpetology classes trapping turtles during those months in areas where red-eared sliders were present.

Red-eared sliders are primarily diurnal, sleeping on the bottom or resting in vegetation on the surface at night. During the day, when not foraging, they are basking—an activity that consumes much of the daytime for all age classes from hatchlings to adults. The turtles may bask out of the water on floating mats of vegetation, muskrat lodges, or partially submerged logs, rocks, or stumps—aerial basking; or they may float on the surface of the water if the water is warm enough—aquatic basking. The species has a high preferred average body temperature of 78° F (25.6° C), with a range of 62.2 to 89.8° F (16.8–32.1° C) (Nutting and Graham 1993), and basking allows them to regulate their body temperature. Basking may also promote digestion, raise the metabolic rate, remove ectoparasites—parasites that live on the outside of another animal—and accelerate egg development in females. In Illinois, Cagle (1950) noted that red-eared slider basking activity peaked during midmorning and midafternoon, and Dreslik and Kuhns (2000) found that early spring basking peaked between 1:00 and 2:00 p.m. They also found that medium-sized sliders basked uniformly throughout the day, while peak basking of smaller individuals occurred at midday.

Home range size varies by habitat type—marsh, pond, river, or lake—but can be extensive. Across the turtle's geographic distribution, home range size has been reported as 1.7 to 809.5 acres (0.7–327.6 ha; Jaeger and Cobb 2012). In

Basking red-eared slider from Linn County. Photo by Kathryn N. Deetz.

a large Tennessee lake, red-eared sliders made longer movements, maintained larger home ranges, and selected deeper areas of the lake than did midland painted turtles (*Chrysemys picta marginata*) (Jaeger and Cobb 2012). Individual home ranges may include multiple bodies of water between which the turtles travel frequently (Cagle 1944a), and these overland trips may be undertaken in search of food, mates, nesting areas, basking sites, or overwintering sites. Movements may also be forced by the drying out of aquatic habitats, at which time the turtles seek permanent water. Though they do not truly estivate, during extreme temperatures—above 87.8° F (31° C)—they may burrow into the mud, enter stumps, or seek shelter in shaded waters (Cagle 1950).

Fifty-one marked red-eared sliders in Michigan exhibited a degree of homing ability, though details of distance traveled, time involved, and environmental conditions were vague (Cagle 1944a). In a study of the related yellow-bellied slider (*T. s. scripta*) (Yeomans 1995), individuals were released on land in an unfamiliar area that had a single nearby source of standing water. In overcast conditions, the turtles oriented randomly; however, in clear conditions, they oriented nonrandomly and in the direction of the nearest water, suggesting that adult pond sliders have a water-finding ability. Yeomans concluded that turtles with the ability to move to another aquatic habitat in a direct and efficient manner should have an advantage because reduced exposure to predators, reduced water stress, and reduced time away from feeding would lower the cost of overland travel.

As fall approaches, the turtles become less active, eventually becoming inactive at water temperatures below 50° F (10° C; Cagle 1950). They overwinter in hollow logs or muskrat dens—18 were found in a single muskrat burrow in southern Illinois (Cagle 1950)—but they may also overwinter lying in the open on the bottom of a pond or lake. The latest Iowa record is from October 7 in Polk County, though this is from an introduced population. The latest record from within the turtles' native Iowa range is from September 10 in Lee County.

THREATS. A wide variety of insects, fish, large frogs, reptiles, birds, and mammals eat red-eared slider eggs and hatchlings. Raccoons and skunks can be major nest predators. However, large adults probably have few predators in Iowa. The turtle has excellent hearing and will slide into the water at the slightest disturbance.

Although largemouth bass have been shown to eat some hatchling turtles (Lundgren et al. 2012), experiments suggest that they may not be a major predator of sliders. In the first experiment, live hatchling yellow-bellied sliders were shown to be swallowed by the bass but then quickly spit out, whereas dead hatchlings with the margin of the carapace removed were swallowed and not spit out

(Semlitsch and Gibbons 1989). The researchers suggested that either the behavior of the hatchlings or their carapace and plastron architecture might prevent them from being swallowed. A later experiment showed similar results: live hatchling red-eared sliders were attacked by largemouth bass but again spit out (Britson and Gutzke 1993). The researchers concluded that when hatchlings claw and bite, they harm or irritate the gill apparatus or digestive tract of the bass, providing defense against predation.

In Iowa, red-eared sliders cannot be trapped or otherwise collected; however, other states do allow trapping, either for personal consumption as food or for exportation as food or pets. Trapping tends to remove larger adults from a population, which can lead to local population declines. At one time, it was estimated that 5 to 10 million red-eared sliders were exported annually around the world (Ernst and Lovich 2009). To accommodate the pet trade, turtle farms were established, but turtles were still removed from the wild to keep up with demand, often to the detriment of natural populations (Cadi and Joly 2004).

The pet turtle trade not only affected turtle populations but also raised human health concerns. From the 1950s to the mid-1970s, hatchling red-eared sliders were popular pets, often sold in dime stores with little plastic ponds containing a small raised island and a plastic palm tree. The young turtles rarely lived long, largely due to improper care, and were often released into neighboring rivers and ponds. We both had them as children; Neil remembers his father releasing his into a river. However, in 1975, the U.S. Food and Drug Administration banned the sale of live turtles under 4 inches (10 cm) straight-line carapace length in the United States due to concern over the transmission of the disease salmonellosis.

Pet turtles as a source of salmonellosis in children was first documented in 1965, and later studies showed that 23 percent of families in which salmonellosis had been diagnosed owned or had kept a pet turtle, compared with only 6 percent of families that had not (Shane et al. 1990). In the early 1970s, Lamm et al. (1972) estimated, based on an annual sale of 15 million hatchling turtles in 1971, that approximately 4 percent of all households in the United States maintained one or more turtles, each of which had an average lifespan of only 2 months. Approximately 280,000 cases of turtle-associated salmonellosis occurred in the United States in 1970 and 1971, with related medical expenses amounting to millions of dollars. Though the number of cases of salmonellosis declined after the 4-inch (10-cm) turtle ban, the threat has not gone away. In September 2023, the Centers for Disease Control announced that public health officials were investigating multiple outbreaks of salmonella linked to small turtles in 18 states, with almost 40 percent of those infected being children younger than 5.

An interesting study by Tucker, Dolan et al. (2008) demonstrated the mixed effects of climatic warming on red-eared sliders on the Illinois side of the Mississippi River adjacent to Iowa. During a warm period between 2001 and 2004, the population of turtles more than doubled because females laid more eggs during the longer nesting seasons. However, the extra clutches were laid when soil temperatures were relatively low; thus, most new individuals were male, resulting in an almost 7 percent decline in females between 1994 and 2006.

FOOD AND FEEDING. *T. s. elegans* is an omnivorous aquatic forager that consumes a wide variety of plants and animals, though its bite force is rather weak, which likely limits the prey it can consume to small soft-bodied animals or carrion. Known foods include algae, bryophytes (mosses and liverworts), a large variety of vascular plants, aquatic invertebrates, fish, and frog eggs, tadpoles, and adults. Other turtles (as determined from the presence of scutes in the stomach), snakes, and small birds are also consumed. When foraging, the turtle moves through shallow water with abundant aquatic vegetation using its head and forelimbs to disturb the plants and flush prey. It is attracted to movement and will chase fish, tadpoles, and frogs. Small prey is eaten whole, while larger prey is held in the jaws and shredded with the front feet.

Sliders are reported to go through a dietary shift as they age. Juveniles are highly insectivorous but shift to progressively more plant food as they grow, though adults still prefer animal food when it is available (Clark and Gibbons 1969; Hart 1983; Parmenter and Avery 1990). The dietary shift corresponds to a habitat shift as the turtles grow—juveniles forage in shallow water and move to deeper water as they get larger. Dreslik (1999) reported somewhat contradictory results in Illinois, where he found that both sexes consumed approximately equal amounts of plant and animal matter, while the diets of juveniles contained over 70 percent plant matter. He also reported that the overall diversity of food items in the diet of red-eared sliders was large, and the diversity of items expanded from early to late seasons. Females consumed more mollusks than males, and males and females consumed more bryozoans—aquatic colonial invertebrate animals—than juveniles. Webb (1961) suggested that the omnivorous diet of the red-eared slider allowed them to exploit the differing availability of foods during fluctuations in water level.

REPRODUCTION. The age and size at which red-eared sliders become mature vary by location, and no geographic pattern in age at maturity is apparent based on the available data (Lewis et al. 2018). In Illinois—likely most representative of Iowa red-eared sliders—males reach sexual maturity at straight-line carapace lengths greater than 4 inches (10 cm), and females become mature at greater than 6.7 inches (17 cm) straight-line carapace length (Cagle 1944b, 1948).

Males produce sperm in the fall and store it over the winter for copulation in the spring. All males in the Drake University Research Collection that were examined had sperm present in August. Vitellogenesis—ovarian follicular enlargement—begins in late summer or early fall with follicles reaching ovulatory size by November or December. Counts of enlarged ovarian follicles in Iowa females in the Drake University Research Collection found 15 enlarged ovarian follicles, ranging from 0.47 to 0.6 inch (1.2–1.5 cm) in diameter, along with a smaller set of 15 follicles ranging from 0.4 to 0.47 inch (1.1–1.2 cm) in diameter, in a Muscatine County female found on August 7. Follicular development continues until ovulation the following May.

Most breeding occurs from early spring to early summer, and fall mating has been observed in some locations. In Illinois, red-eared sliders were observed courting from March through June and again from late September until overwintering (Cagle 1950). During courtship, the male approaches the female from the front with his forelegs together, palms out, and drums his greatly elongated foreclaws on her head. If receptive, the female sinks to the bottom, allowing the male to follow from behind. Once aligned over the back of the female, the male clasps her carapace with all four feet, bends his head downward to touch hers, and curls his tail around to align the vents. During copulation, the male releases his forelimb grip, pulls his head into his shell, and floats to an almost vertical position. Mating may last up to 15 minutes. Several males may simultaneously court the same female, with larger males having an advantage. Melanistic males court more frequently than nonmelanistic males.

Nesting occurs from April to mid-July—likely peaking in May and June—in unshaded nonmuddy areas near the water. When searching for a suitable nest site, the female may dig several test holes before choosing a site. In Louisa and Lee Counties, Jim observed nesting during the last 2 weeks of June approximately 50 feet (15.3 m) from oxbows and inlets of the Mississippi River. In Clark County, Missouri, immediately south of Lee County, Iowa, nesting has been observed in late May and early June (Briggler and Johnson 2021). Using her hind limbs, the female excavates with alternating scoops a flask-shaped cavity up to 4.5 inches (11.5 cm) deep. If the substrate is hard, she may empty her bladder, presumably to soften the soil. Once the eggs have been laid, the female uses her hind legs to push the soil back into the hole to cover the eggs, kneading the soil with her hind limbs and moistening the nest with water from her bladder. Cagle (1950) described the mixture of wet soil and debris capping the nest as resembling "a mud ball thrown

Hatchling red-eared slider from Louisa County. Photo by Ryan D. Rasmussen.

forcibly against the ground." Females may return to the same nesting area year after year.

Clutch size is positively correlated with female size, with larger females producing more eggs. Across their range, females lay 1 to 2 clutches per year each containing 1 to 30 eggs (average = 10). The eggs are white, oval, 1 to 1.6 inches (2.6–4 cm) long, and granular. In southern Illinois, clutch size averaged 9.3 eggs (Cagle 1950), while in west-central Illinois clutch size ranged between 12.5 and 15.1 eggs (Tucker, Dolan et al. 1998). A nest containing 19 eggs was found just south of the Iowa border in Clark County, Missouri (Briggler and Johnson 2021). A Louisa County female in the Drake University Research Collection found on May 29 contained 6 eggs.

The eggs hatch in late August or September with the hatchlings overwintering in the nest and emerging the following spring. Six eggs incubated in Jim's lab at Drake University hatched on August 19, and Jim and Neil have both captured hatchling red-eared sliders in drift fences in Louisa County in late May and June as they were leaving nests and moving to the water.

The bright green hatchlings are nearly round, have a dorsal keel, and range from 1 to 1.6 inches (2.6–4 cm) straight-line carapace length. Similar to most turtles, red-eared sliders lack sex chromosomes, and the sex of hatchlings is determined by the temperature of the nest—temperature-dependent sex determination. Males are produced at incubation temperatures between 72.5 and 80.6° F (22.5–27° C);

females are produced at 86° F (30° C). Hatchling survivorship is correlated with straight-line carapace length and body mass, both of which are correlated with soil moisture that retards water loss (Janzen et al. 2000a, 2000b; Filoramo and Janzen 2002).

LONGEVITY. A juvenile survived in captivity at least 37 years, 9 months, and 10 days. Gibbons and Semlitsch (1982) estimated that fewer than 1 percent of wild pond sliders reach an age of 20 years and probably have a maximum lifespan of 30 years. Gibbons (1987) reported recaptures of wild pond sliders with a minimum age of 25 years and considered the maximum lifespan to be 35 years.

FAMILY KINOSTERNIDAE

MUD AND MUSK TURTLES

The family Kinosternidae includes four genera and approximately thirty species from Canada to South America. Two genera, *Kinosternon* and *Sternotherus*, and two species, the yellow mud turtle and the eastern musk turtle, occur in Iowa. Kinosternid turtles appear to have evolved in Mexico or Central America and later moved northward; the earliest fossil in North America dates to at least the Eocene epoch 55 to 34 million years before the present. Phylogenetically, the family is most closely related to the Chelydridae.

Members of the family are relatively small with short tails, dull coloration, barbels on the chins—small whisker-like structures used for smelling, from the Latin *barbula* meaning "little beard"—non-paddle-like limbs, jointed toes, and four or five claws. They have only ten or eleven scutes on their single- or double-hinged plastrons. Eastern musk turtles have small plastrons with only the anterior portion movable; yellow mud turtles have large plastrons with both the anterior and posterior portions movable. Kinosternids have musk glands on the plastral bridge that release smelly secretions when the turtle is handled, and they can be highly aquatic or mostly terrestrial, depending on the species, time of year, and location.

YELLOW MUD TURTLE

Kinosternon flavescens (Agassiz) 1857

Yellow mud turtle from Louisa County. Photo by Don Becker.

IOWA STATUS. Endangered; species of greatest conservation need.

HISTORICAL SUMMARY. The yellow mud turtle has also been called the Illinois mud turtle in the Midwest. The first confirmed Iowa specimen was collected as roadkill by Reeve Bailey and a coworker in 1939 south of Muscatine in Muscatine County (University of Michigan Museum of Zoology), though at the time Bailey believed it to be an eastern mud turtle (*K. subrubrum*) and a probable introduction since the species was not previously known from Iowa. Carlander and Moorman (1951b) mentioned *K. subrubrum* in their discussion of Iowa's pond and river turtles, stating, "It is possible that this species should not be considered as an Iowa species since it has been collected only once and since its normal range is mostly to the south." Dodge and Miller (1955) reported collecting 15 yellow mud turtles—males, females, adults, and juveniles—at two locations in Muscatine County, which they believed to be the first specimens collected in Iowa, presumably unaware of Bailey's earlier collection from the same location. Jim

Christiansen, during his 37 years at Drake University, conducted the first detailed and long-term studies of yellow mud turtles in Iowa, mostly centered at Big Sand Mound Nature Preserve—private land in Louisa and Muscatine Counties jointly owned and managed by Bayer and MidAmerican Energy Company.

John Iverson and Jim (2023) provide a complete review of the taxonomy of the yellow mud turtle. Three subspecies of *Kinosternon flavescens* were previously recognized. Smith (1951) suggested that the Midwest populations be given distinct subspecies status and the name Illinois mud turtle (*K. f. spooneri*). However, morphological and molecular studies (Houseal et al. 1982; Berry and Berry 1984) found no evidence supporting separate subspecies status for Midwest mud turtles, and as a result only the yellow mud turtle is currently recognized in Iowa.

DESCRIPTION. The yellow mud turtle is a drab little turtle with an oval and high-domed flattened carapace that is yellow-brown, tan, dark brown, or black with dark seams between the scutes. In Iowa, the turtles are noticeably darker than are those to the south and west but become lighter after about age 20. The carapace has 11 marginal scutes on each side that are tan or black above and yellow below, and the ninth and tenth marginals are distinctly higher than the eighth marginal scute. The yellow to brown plastron lacks exposed skin between the

Plastron of a yellow mud turtle from Louisa County. Note the triangular pectoral scutes.

plastral scutes and has 11 scutes, triangular pectoral scutes, a prominent notch between the posterior scutes, and a double hinge—*Kinosternon* comes from the Greek *kineo* meaning "move" and *sternon* meaning "chest" or "breast" in reference to the turtle's ability to move its plastron; *flavescens* comes from the Latin *flavus* meaning "yellow" and *-escens* meaning "becoming" in reference to the color on the underside of the jaw and throat. A series of growth rings are visible on the plastron and may be used to estimate the age of the turtle, although they become unreliable after about age 12. The skin is gray to black dorsally and cream to yellow ventrally, 2 pairs of prominent barbels are present under the chin, and the toes are webbed. The barbels and the front half of the lower jaw are yellow. Hatchlings are brown with black spots located at the posterior margin of each of the large scutes on the carapace and yellow on the edges of the marginals.

Males have short concave plastrons, moderately hooked upper jaws, 2 small patches of rough scales on the inner surface of each hind leg, and long thick tails at least half as long as the turtle is wide with a prominent spine at the tip. Females have flat plastrons, weakly hooked upper jaws, much smaller tails with a less prominent spine, and no patches of rough scales on the inner surface of the hind legs.

SIZE. Yellow mud turtles generally range from 4 to 5 inches (10–12.5 cm) straight-line carapace length. Iowa adults typically range from 3.5 to 5.7 inches (9–14.5 cm) straight-line carapace length. The largest specimen in the Drake University Research Collection is from Louisa County measuring 5.4 inches (13.7 cm) straight-line carapace length. The record straight-line carapace length reported for the species is 6.6 inches (16.8 cm).

SIMILAR SPECIES. Three other Iowa turtles have hinged plastrons—ornate box turtles, Blanding's turtles, and eastern musk turtles—but all have only a single hinge compared to the double hinge of the yellow mud turtle. The carapace of the ornate box turtle has a dark ground color, light to bold yellow lines on each scale that radiate downward, and a similarly patterned plastron. Blanding's turtles have a dark brown to black carapace with white to yellow speckling or irregular radiating lines and a bright yellow neck and chin. The eastern musk turtle is the most similar, but it can be distinguished from the yellow mud turtle by the presence of a single poorly developed plastral hinge that is only barely movable, square pectoral scutes, broad areas of white instead of dark connective tissue separating the plastral scutes, and a ninth marginal scute that is the same size as the eighth marginal scute.

DISTRIBUTION. Yellow mud turtles have a relatively large geographic range that includes isolated populations in western Illinois, southeast Iowa, northeast, west-central, and southwest Missouri, and northern Nebraska; however, the Illinois, Missouri, and Iowa populations are greatly reduced or extirpated (Christiansen et al. 2012). More contiguous populations extend south from southern Nebraska through Kansas, Oklahoma, and Texas into Chihuahua and Veracruz, Mexico, and west into eastern Colorado, New Mexico, and southeast Arizona. Iverson (1979) believed that the species evolved in the semi-arid grasslands of southwest North America and entered Iowa, Illinois, and Missouri during the xerothermic period—a warm dry postglacial interval approximately 7,000 to 5,000 years before the present. The high temperatures and dry conditions favored the formation of prairies across much of Iowa. This period probably encouraged reptile migration; the distribution of Iowa's more arid-adapted reptiles, like the yellow mud turtle, can be explained in part by former connections to populations that occurred in grasslands and similar xeric habitats to the west and southwest of the state.

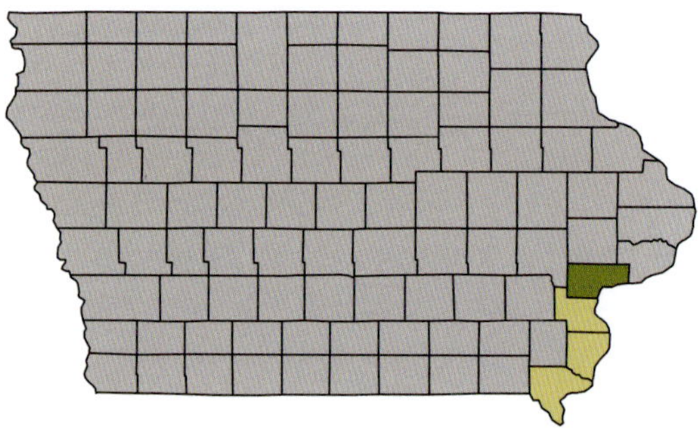

Yellow mud turtle distribution in Iowa.
Yellow = post-1960
Green = both pre- and post-1960

In the mid-1940s, the species had been documented in just one county, Muscatine; at present, records exist for just 4 of Iowa's 99 counties, though it is probable that the turtles exist only in Muscatine and Louisa Counties. By the early 1980s, Jim (1981) had concluded that they had declined in Iowa due to the loss of sandy habitat adjacent to semipermanent wetlands in the Mississippi River floodplain in southeast Iowa. In 1990, Jim and his coworkers (1990) reported that only one known Iowa population of mud turtles was large enough to be considered viable, and by the late 1990s he (1998) stated that extensive trapping efforts in areas where they had previously occurred suggested that no other dense population existed comparable to the Big Sand Mound Nature Preserve population. A search for the turtle by Neil and Jim (2011) within the formerly known Iowa range did not produce any outside of Big Sand Mound Nature Preserve. Monitoring of the Big Sand Mound Nature Preserve population by Joshua Otten, Donald Becker, and Jim Scharosch continues. Their work has shown that the population there continues to steadily decline with little if any successful reproduction, resulting in the yellow mud turtle being one of the most imperiled reptiles in Iowa.

HABITAT. Yellow mud turtles are adapted to living in arid environments where water is available only for short periods of time. They evolved on the Great Plains and likely coevolved with the American bison, whose wallowing behavior created depressions, many of which became seasonal wetlands that although temporary provided suitable habitat and food for the turtles (Iverson and Christiansen 2023). Across their range, yellow mud turtles inhabit swamps, sloughs, and other wetlands, rivers, creeks, ponds, cattle tanks, open woodlands, and even deserts. The aquatic habitats are characterized by shallow water, soft bottoms, and aquatic vegetation. A critical component is the availability of soils consisting of nearly pure sand in the uplands surrounding the aquatic habitats, which are essential for burrowing during summer estivation and during winter at the northern edge of their range to get below the frost line.

In Iowa, the turtle is found exclusively in the sand deposits of the Mississippi Alluvial Plain (Prior 1991). Extensive surveys by Bickham et al. (1984) and by Neil and Jim (2011) showed that yellow mud turtles in Iowa are greatly limited by the availability of fine clean sand. Big Sand Mound Nature Preserve, the area adjacent to the Mississippi River that supports the largest and possibly the only remaining population of yellow mud turtles in the state, is a large sand deposit rising 65.6 to 98.4 feet (20–30 m) above the floodplain (Christiansen et al. 1985). The area between the sand dunes and the river consists of floodplain woodlands and two large oxbow lakes interspersed with smaller temporary and permanent ponds. Most stunning are the open shifting sand dunes—certainly unique in Iowa—sparsely vegetated with arid-adapted grasses and other plants, including prickly pear cactus. The combination of upland shifting sandy soils and adjacent shallow wetlands creates ideal habitat for the yellow mud turtle.

The research of Jim and his many coworkers at Big Sand Mound Nature Preserve has shown that ideal aquatic feeding habitat for yellow mud turtles consists of shallow ephemeral—lasting for a very short time—wetlands, which reduce the numbers of both competitors and predators. When the wetlands are filled by spring floods or rising water tables to a depth of 1.6 to 6.6 feet (0.5–2 m), with most averaging less than 3.3 feet (1 m) deep, they are rapidly invaded by yellow mud turtles emerging from their overwintering locations. The wetlands contain

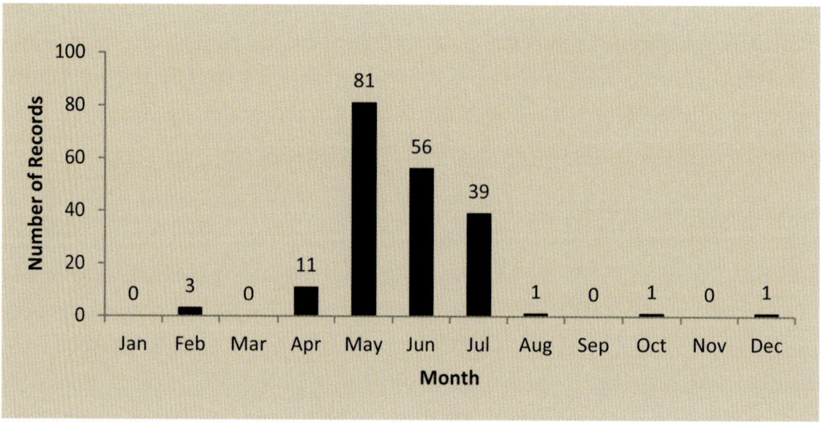

Collection/observation dates of 193 yellow mud turtles in Iowa.

a mix of emergent, submergent, and floating vegetation. Predatory fish, mainly sunfish, are introduced into the wetlands by spring floods, though many die later in the season as the wetlands dry up and become oxygen depleted or as a result of winterkill.

BEHAVIOR. Yellow mud turtles are usually docile when handled, though some may bite if they have the opportunity. Male-male combat, consisting of mounting, genital biting, and chasing and biting, has been observed and is assumed to be a territorial defense of the aquatic areas that provide access to food and females (Lardie 1983). The stronger and more dominant male is usually the aggressor.

The collection/observation dates of 193 Iowa yellow mud turtles show that the species is active from mid-April through July in the state; data from over 50 years of monitoring at Big Sand Mound Nature Preserve show a similar pattern (Jim Christiansen and Joshua Otten, personal communication, 2023). The earliest and latest records are three found dead on February 20 and one found dead on December 2. These may have been winterkills that attempted aquatic overwintering. The August and October specimens are believed to have returned to water to rehydrate and were killed by raccoons. Yellow mud turtles have one of the shortest activity periods of any North American turtle, spending most of the late summer, fall, and winter buried in soil. The studies of Jim, Neil, and their coworkers at Big Sand Mound Nature Preserve along with Jim's 37 years of personal observations of the species in Iowa were used for the information below except as noted.

Most adults and juveniles emerge from overwintering by the last third of April,

Yellow mud turtle from Louisa County beginning to bury itself. Photo by Don Becker.

when they move into the closest shallow water to rehydrate. Hatchlings emerge later—from late April through the third week of June. Newly emerged turtles appear thin and dehydrated, and often the area under the shell around the legs is packed with sand. The turtles may be somewhat lethargic at this time and may fall prey to raccoons if they do not quickly reach the water. They continue to emerge well into May and remain in the water, where they feed, mate, and put on weight. By mid-June, females begin leaving the water to nest in nearby sand dunes, which continues into July. Most females do not return to the water but instead remain in the dunes until the following spring. Sometime in early July, males begin leaving the water, and by the third week of July nearly all the mud turtles have left the water and returned to the dunes, where they estivate terrestrially for the rest of the summer, resulting in an aquatic activity period of approximately 80 to 90 days in Iowa (Iverson and Christiansen 2023). In some years, a few individuals may return to the water for a short bout of fall feeding.

Yellow mud turtles spend much of the year underground, either estivating or overwintering. In Texas, Rose (1980) reported that they can estivate for as long as 2 years, and Joshua Otten, Donald Becker, and Jim Scharosch have evidence of a male at Big Sand Mound Nature Preserve that did not emerge for 672 consecutive days (Donald Becker, personal communication, 2023). Estivation sites in Iowa are in upland sand dunes adjacent to wetlands; once the turtles leave the water, they burrow into the sand and are rarely seen again until the following spring, though they often relocate multiple times during the estivation season. In Iowa, Cooper (1975) observed movement of some individuals that had left the water, especially females that burrowed several times before nesting. All 14 radio-tracked turtles studied by Jim and his coworkers (1985) changed locations in the dunes during July and August, most of them multiple times. One female rehydrated in a pond for 10 days in August before returning to the dunes. In Illinois, 3 radio-tracked females used an average of 9 different burrows in one season, and 5 males used an average of 3.4 burrows; estivation depth was reported to be 2 to 9.8 inches (5–25 cm; Tuma 2006). In Nebraska, yellow mud turtles often relocated if they were disturbed even slightly (Iverson and Christiansen 2023), and estivation relocation has been shown to be stimulated by rainfall events (Christiansen et al. 1985; Tuma 2006).

These turtles are wonderfully adapted to surviving extended periods of estivation. Long (1985) reported that the turtle has the highest level of body fat compared to other turtle species and postulated that this may be an adaptation of fat loading to survive in a fluctuating environment. Oxygen consumption and carbon dioxide exhalation are also reduced, and the turtles become anaerobic—existing in the absence of oxygen—even after only 2 weeks underground (Seidel 1978). Their skin also thickens to reduce water loss (Seidel and Reynolds 1980). Individual turtles can lose up to a third of their weight during estivation, mainly from water loss; however, they are able to quickly rehydrate once they enter the water (Seidel 1978).

As fall approaches, the turtles begin burrowing deeper to stay ahead of the advancing frost line. They are not freeze-tolerant and must get below the frost line to survive. Most individuals overwinter terrestrially, and very few Iowa adults remain in the water over the winter. At Big Sand Mound Nature Preserve, yellow mud turtles overwinter 328 to 1,640 feet (100–500 m) from the water. In Illinois, overwintering depth was reported to be 42.1 to 49.2 inches (107–125 cm; Tuma 1993a, 2006). Terrestrial overwintering limits winterkill, which is more likely with aquatic overwintering in the shallow ephemeral wetlands used by this species. The consequences of aquatic overwintering in Iowa became known with the discovery of 8 presumably winterkilled yellow mud turtles, along with approximately 50 other turtles, at Big Sand Mound Nature Preserve by Murphy and Corn (1977), and the selective pressures against aquatic overwintering were explored by Christiansen and Bickham (1989).

THREATS. Yellow mud turtle eggs and hatchlings are eaten by a variety of predators, including large fish, bullfrogs—though Neil examined stomach contents of bullfrogs at Big Sand Mound Nature Preserve and did not find any turtle remnants—snapping turtles, watersnakes, raccoons, and striped skunks. Plains hog-nosed snakes, which also occur at Big Sand Mound Nature Preserve, have been reported to use the scent trails of female yellow mud turtles to find their nests (Iverson 1990) and have been observed consuming yellow mud turtle eggs in Illinois (Tuma 2006) and in Nebraska—where they were the primary predators, taking 48.9 percent of the monitored eggs (Iverson 1991).

Although the plastron does have hinges that allow it to close, the carapace does a poor job of protecting the underside, subjecting the turtle to predation by a variety of mammals. During his 37 years of studying yellow mud turtles in Muscatine and Louisa Counties, Jim recalls that raccoons, along with striped skunks and possibly coyotes and opossums, took the largest toll on this turtle. The turtles are often found by raccoons as they move between their terrestrial overwintering sites and the water. The turtle is usually carried to the water's edge, where the raccoon pulls its legs out of its shell—dragging the internal organs out with them—and consumes the soft parts, leaving an empty shell. Predation is the worst during spring emergence but also occurs when the turtles leave the water to nest or estivate.

The impact of raccoons on Big Sand Mound Nature Preserve turtle populations was demonstrated after raccoons were heavily trapped at the site in early 1979. Jim and Benny Gallaway (1984) reported that only 14 hatchlings of all turtle species were captured in 1978; however, between 75 and 80 were captured during each of the following 3 years. Raccoons began recolonizing the area in 1981, and by 1982 they were again abundant, which resulted in only 30 captured hatchlings in 1982. Juvenile yellow mud turtles made up an average of 21.1 percent of the trapped population prior to raccoon removal and 41.7 percent after their removal.

Like most turtles, yellow mud turtles are susceptible to both parasites and diseases. Leeches are commonly found on individuals outside of Iowa, although they were not commonly seen in Iowa despite being present in large numbers in the ponds used by the turtles (Christiansen et al. 1985). Jim and his coworkers (2021) described a shell disease affecting yellow mud turtles at Big Sand Mound Nature Preserve that involved the loss of coloration of the scutes on the carapace followed by the loss of portions of the scute and exposure of the underlying bone; however, the cause was undetermined. The number of turtles with the shell disease has appeared to decline since the 1980s, and the loss of color does not seem to affect the survival of individuals. The condition, at least in the earliest stages, affects females more severely than males and worsens with age in both sexes.

The yellow mud turtle's primary threat is habitat modification through the loss of ephemeral and permanent wetlands due to agriculture, urbanization, droughts, draining, ditching, water table declines resulting from groundwater pumping for agriculture, industry, and housing, and woody encroachment of adjacent sand dunes. Because the turtle's year is almost equally divided between terrestrial and aquatic habitats, both sandy uplands and adjacent predictable bodies of water are essential. If the turtle cannot burrow into the sand, it has no place to lay eggs, estivate during the hottest part of the year, or survive freezing winter temperatures. Without water in the spring, the turtles cannot feed and rehydrate.

The area around Big Sand Mound Nature Preserve has experienced both the loss of sand prairies to development and the loss of wetlands to groundwater withdrawals coupled with the prevention of flooding due to levees constructed along the Mississippi River (Bernstein and Christiansen 2011; Christiansen et al. 2012). At Big Sand Mound Nature Preserve, the shifting sand dunes favored by arid-adapted species like yellow mud turtles are being encroached upon by woody and herbaceous plants that stabilize the soil, thereby reducing the wind erosion essential for maintenance of the dunes.

Climatic changes have the potential to affect the turtles as well. Because of their reliance on wetlands in the spring to feed and hydrate, yellow mud turtles at Big Sand Mound Nature Preserve were potentially stressed when a major oxbow dried up during the drought of 1988. In response, an artificial depression 7 feet (2.1 m) deep was excavated to hold water during dry periods, and both yellow mud turtles and western painted turtles colonized the wetland in the same proportions (Watts and Christiansen 1989). While young yellow mud turtles used the artificial wetland, their shells were softer than normal, a consequence attributed to a calcium deficiency resulting from low snail populations in the newly created depression. Subsequent studies have shown that the wetland now largely supports sunfish, yellow perch, and western painted turtles (Neil Bernstein and Donald Becker, personal communication, 2023).

In the northern Great Plains, yellow mud turtle reproduction has been shown to be sensitive to temperature changes. Following the 1991 eruption of Mount Pinatubo in the Philippines, global temperatures were reduced for two years. The record-low temperatures in 1992 and 1993 resulted in virtually no recruitment of new yellow mud turtle hatchlings into a Nebraska population in 1993 and 1994 (Iverson 2022). The low summer temperatures prolonged incubation times to such an extent that hatching was nearly impossible before winter mortality became a problem. Thirty years later, a gap still exists in the age demographics of the population. At Big Sand Mound Nature Preserve, new hatchlings captured numbered 168 in 1988, 41 in 1991, and 5 in 1994. Similarly, adult captures declined from 302 in 1991 to 159 in 1994 to 174 in 1997 and continued to decline thereafter. While a direct causal effect of the volcanic eruption cannot be drawn for Iowa, interestingly the decline in hatchling captures in the state began after 1991. In contrast, modeling of how climate change will affect Kinosternid turtles as the atmosphere continues to warm showed that climatic conditions (not habitat) favorable to yellow mud turtles was expected to increase between 2050 and 2070 in the Midwest (Butler et al. 2016).

While once more abundant, the remaining Big Sand Mound Nature Preserve population is small with little reproduction (Christiansen et al. 2012; Jim, Neil, and Joshua Otten, personal observation, 2023). At present, the population is decreasing and aging, has a low percentage of females and low reproductive viability, and is probably losing genetic variation. Because populations in Illinois and Missouri are also declining, yellow mud turtles in the Midwest are in danger of extirpation.

FOOD AND FEEDING. *K. flavescens* is an opportunistic, primarily carnivorous feeder that eats a variety of invertebrates, amphibians and their larvae, fish, birds, algae, duckweed, and plant debris. In Iowa, Jim and his coworkers (1985) found through stomach content analysis a predominance of snails, duckweed, beetles—mostly aquatic—seedpods, and leeches. Duckweed was present in 61 percent of the stomachs, but like other plant material it may have been ingested incidentally. Examination of feces revealed crayfish parts and unidentifiable bones. Feeding mainly ceases in July during terrestrial estivation, though some individuals may consume underground invertebrates while buried.

The turtles are able to shift their diet to take advantage of different prey as heavy rain or drought affects the abundance of different food items. Fish, crayfish, and microcrustaceans were the most abundant prey found in stomachs during the May to July feeding period; when those items were not abundant, the turtles fed primarily on snails, aquatic and some terrestrial beetles, and fly, dragonfly, and damselfly larvae. Early May feeding included mayflies, bees and wasps, and butterflies and moths, but those items were rarely seen in the diet in later months. Jim recalls that during the extensive and prolonged flooding in 1993 and 1994, mud turtles were seen moving under the clear water in flooded areas and gorging themselves on earthworms. During this time, the turtles avoided baited traps because they were so stuffed with earthworms, and if they tried to pull their legs into their shells, they regurgitated the worms.

REPRODUCTION. Males reach sexual maturity in 5 to 6 years at straight-line carapace lengths of 3.1 to 3.5 inches (8–9 cm), while females mature at 3.1 to 4.9 inches

(8–12.5 cm) in 6 to 7 years in southwest populations and in 9 to 15 years in northern populations (Iverson and Christiansen 2023). Examination of 28 Iowa adult female specimens in the Drake University Research Collection found enlarged ovarian follicles in an estimated 10-year-old at a straight-line carapace length of 3.6 inches (9.1 cm) and an estimated 11-year-old at a straight-line carapace length of 3.5 inches (9 cm), while an estimated 12-year-old at a straight-line carapace length of 3.7 inches (9.5 cm) was immature and contained no enlarged ovarian follicles or corpora lutea, suggesting that Iowa females mature at around 10 to 12 years of age and 3.5 to 3.7 inches (9–9.5 cm) straight-line carapace length.

Mating usually occurs in the water soon after spring emergence. In Iowa, it has been observed as early as May 6 even though the water temperature was only 64.4° F (18° C) at the time (Christiansen et al. 1985). Courtship and mating were described by Mahmoud (1967) and Lardie (1975a, 1975b, 1978) and include three phases—tactile, mounting and intromission, and biting and rubbing. During the tactile phase, the male approaches the female from behind with his neck outstretched, smelling or touching her tail to confirm the sex, and may push or bite at her tail and head. If unreceptive, the female moves away; however, the male may chase, nudge, and bite her. If she is receptive, the male approaches her from behind or the side and uses the larger second and third digits of his forelimbs to cling to her first and second marginal scutes and the large claws on the first three digits of his hind feet to clasp her ninth and tenth marginals, allowing him to grasp her carapace as he mounts. The male flexes a rear knee and holds the female's tail between the 2 small patches of rough scales on the insides of his hind legs, maintaining this position by looping his tail so that the spine on the end of his tail contacts the female near her vent. As the vents come together, the male inserts his penis—intromission—followed by rubbing and biting. Mating may last 10 minutes to 3 hours.

Maximum sperm production in Iowa males occurs in August, about a month earlier than in southern populations (Christiansen et al. 1984), and sperm is stored until the following spring. Unlike most North American freshwater turtles, the ovarian cycle of yellow mud turtles is continuous rather than cyclical (Iverson and Christiansen 2023). By the time the females emerge from overwintering in the spring, they already have a set of enlarged ovarian follicles for that year's clutch, as well as a second set of enlarged follicles that represent the following year's clutch. Jim and his coworkers (1984) estimated ovulation to occur in late May or early June in Iowa.

At Big Sand Mound Nature Preserve, females leave their ponds to lay eggs from mid-June through mid-July. They move nonstop directly to sandy areas with little vegetation and usually bury themselves on south-facing slopes. They may try several areas before finally burying themselves about 5 inches (12.5 cm) deep by using both their forelimbs and their hind feet to "swim" through the sand. What happens next is one of the fascinating aspects of yellow mud turtle reproduction—they nest while completely buried underground. After burying herself in the sand, the female excavates a nest chamber another 1.5 to 4 inches (3.9–10 cm) below her by using her hind limbs, then lays her eggs and urinates into the chamber. She then remains buried over the nest cavity for as little as a day to over 38 days—and some females may not leave the burrow until they return to the water the next spring or the spring of the following year—spending 2 years underground. This nest attendance behavior is thought to reduce nest predation, though depredated yellow mud turtle nests have been found at Big Sand Mound Nature Preserve.

In Illinois, some females deposit eggs in two sites (Tuma 1993b). While females south of Iowa may lay two clutches a year, only one clutch per year is likely laid in Iowa. In Nebraska, about 75 percent of females laid eggs during a given year (Iverson 1991). Yellow mud turtles have temperature-dependent sex determination—with lower incubation temperatures producing mostly males and higher temperatures producing females (Iverson and Christiansen 2023). Though this is not conclusive, Vogt et al. (1982) believed that the results of their

Hatchling yellow mud turtle from Louisa County. Photo by Jeff Faircloth.

work suggested a moderate incidence of females at lower temperatures similar to the pattern seen with the eastern musk turtle.

Across the yellow mud turtles' range, clutch size varies from 1 to 10 eggs, with most nests containing 4 to 6 eggs (Iverson and Christiansen 2023). The eggs are white, elliptical, 1 to 1.2 inches (2.6–3.1 cm) long, and hard-shelled. Clutch size tends to increase with body size, but egg size does not. Christiansen et al. (1984) estimated 5.3 eggs per clutch in Iowa. Counts of enlarged ovarian follicles and shelled oviductal eggs in female specimens in the Drake University Research Collection or reported in nests (Christiansen et al. 1984; Jim, personal communication, 2023) resulted in average clutch sizes of 4.6 eggs (range 2–8; n = 20), 6 eggs (range 6; n = 2), and 4.7 eggs (range 4–6; n = 3). Smith (1961) reported a clutch size of 3.5 eggs for a captive female in Illinois, and Iverson (1991) reported the average clutch size to be 6.5 eggs in western Nebraska. Regardless of the method used to estimate the average Iowa clutch size, on average the estimates are 20.6 percent lower (range 7.4–29.1%) than those reported for western Nebraska.

Eggs hatch in the fall after an average 103- to 105-day incubation period. A clutch of 4 eggs hatched in Jim's laboratory at Drake University on September 15. The hatchlings, at a straight-line carapace length of 0.7 to 1.2 inches (1.7–3.1 cm), stay in the nest, digging down below the frost line for the winter, then emerging the following spring. The long incubation time results in the failure of eggs to hatch in colder years, suggesting that temperatures in the active season limit the northern distribution of the species (Iverson 2022).

LONGEVITY. A wild-caught adult survived in captivity an additional 10 years, 4 months, and 25 days. In western Nebraska, these turtles frequently live beyond 50 years, and recapture data suggest that they may live beyond 60 years (Hedrick and Iverson 2017).

EASTERN MUSK TURTLE

Sternotherus odoratus (Latreille, in Sonnini and Latreille) 1801

Eastern musk turtle from Muscatine County.

IOWA STATUS. Threatened; species of greatest conservation need.

HISTORICAL SUMMARY. The eastern musk turtle has also been called the common musk turtle, stinking Jim, and stinkpot. The name common musk turtle is no longer used to avoid suggesting abundance rather than the species' broad geographic range. Stinkpot is likely the most frequently used common name and the one used by most Iowans. Somes (1911) noted the species' occurrence in the state as uncertain. Dodge (1956a) reported 3 females found in Muscatine County, which he believed to be the first specimens collected in the state. However, an eastern musk turtle specimen in the Field Museum of Natural History appears to have been collected much earlier. No date is provided, but other Iowa specimens in the Field Museum attributed to the same collector are dated to the late 1880s, suggesting that the specimen is from the same time period. The location given for the specimen is Ames in Story County with the note, "Story County was added after the original cataloging." The Ames location is certainly not correct, because musk turtles are not known from central Iowa. The first and only detailed study of eastern musk turtles in Iowa was done by Kristi Sutton (1994), one of Jim Christiansen's graduate students and Terry's office mate at Drake.

DESCRIPTION. Eastern musk turtles have a narrow and high-domed light brown to black unpatterned carapace with the ninth marginal scute the same size as the eighth marginal scute. Young turtles have a prominent vertebral keel that is lost as the turtles age and is absent in adults. The small plastron is yellow to brown with 11 scutes, a single well-developed gular scute, white skin showing between the seams of the scutes, and 1 poorly developed hinge—*Sternotherus* is from the Greek *sterno* meaning "chest" and *thairos* meaning "hinged." The pectoral scutes are square—never triangular—and the hinge may be immovable on some individuals. The skin is gray to black with 2 yellow or white stripes on each side of the head and neck beginning at the snout and passing above and below the eyes. On some individuals, the stripes may dissipate into a row of light cream-colored tubercles, or the stripes may be absent altogether. The head is relatively large in relation to the body with a distinctly pointed snout that projects beyond the lower jaw. The beak of the upper jaw usually has 2 downward projections, while the lower jaw has a single sharp upward point. The underside of the chin has 2 stout barbels; 2 additional barbels are located on the neck. Hatchlings are black with a light spot on each marginal scute, a prominent vertebral keel, and a smaller keel on each side. Juveniles often have a pattern of scattered spots or radiating dark streaks on the carapace.

The sexes are dimorphic, with males having longer and thicker tails ending in a blunt hard tip, more exposed white skin on the plastron, 2 small patches of rough scales on the inner surface of each hind leg, and a straight-line plastron length averaging 70 percent of the straight-line carapace length. Females have shorter tails, no rough scales on the inside of the hind legs, a reduced amount of exposed white skin on the plastron, and a straight-line plastron length averaging 75 percent of the straight-line carapace length. Although the sample size is small, examination of sexual dimorphism in 14 specimens (4 males; 10 females) in the Drake University Research Collection found that straight-line plastron length as a proportion of straight-line carapace length differed significantly between males and females (males 70%; females 78%), with females having a longer plastron in proportion to their length.

SIZE. Eastern musk turtles generally range from 2 to 4.5 inches (5.1–11.4 cm) straight-line carapace length. The largest specimen in the Drake University Research Collection is a female from Muscatine County measuring 4.5 inches (11.5

Plastron of an eastern musk turtle from Muscatine County showing the extensive white skin between the plastral scutes and the square pectoral scutes. Photo by Ryan D. Rasmussen.

cm) straight-line carapace length. The maximum straight-line carapace length reported for the species is 5.9 inches (15 cm).

SIMILAR SPECIES. The yellow mud turtle is the most similar but can be distinguished from the eastern musk turtle by its large plastron with 2 freely movable hinges and triangular pectoral scutes, lack of exposed skin between the plastral scutes, and ninth and tenth marginal scutes that are higher than the others.

DISTRIBUTION. Eastern musk turtles are widely distributed throughout eastern North America from southern Ontario and New England south to Florida—excluding the higher elevations of the Appalachians—and west to extreme eastern Iowa, southern Missouri, eastern Kansas and Oklahoma, and eastern and central Texas. Iowa is at the northwest edge of the species' range, with scattered populations found along the Mississippi River and along the lower Cedar and Iowa Rivers in southeast Iowa. In the mid-1940s, Reeve Bailey and others were not aware of the existence of the species in the state (see the discussion about the Field Museum of Natural History specimen above). At present, records exist from only 6 of Iowa's 99 counties, with some formerly stable populations in southeast Iowa showing signs of decline.

Eastern musk turtle distribution in Iowa.

Yellow=post-1960

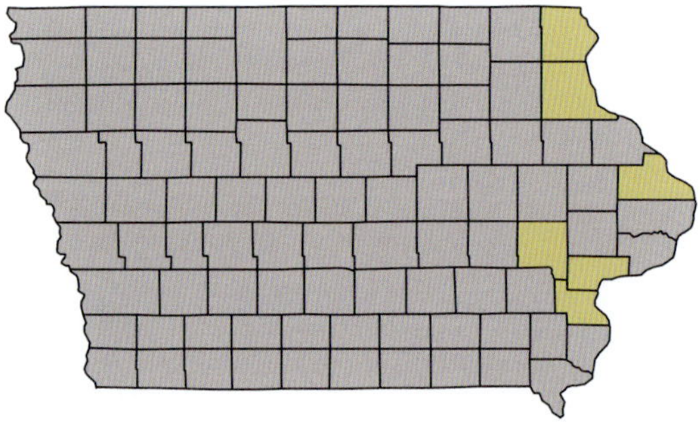

HABITAT. Stinkpots inhabit wetlands, streams, and ponds with a slow current and soft bottom. Lakes, ponds, and their shallow-water edges are preferred over fast-flowing rivers in Illinois (Dreslik and Phillips 2005). In southeast Iowa, backwaters, oxbows, slow streams, and wet road ditches in the floodplains of the lower Iowa and Cedar Rivers are used; in northeast Iowa, the turtle has been found in the Mississippi River. The females reported by Dodge (1956a) in Muscatine County were found in a spring-fed ditch with deep clear water and abundant aquatic vegetation. Equally important as aquatic habitat is the presence of suitable

Eastern musk turtle habitat in Muscatine County. Photo by James L. Christiansen.

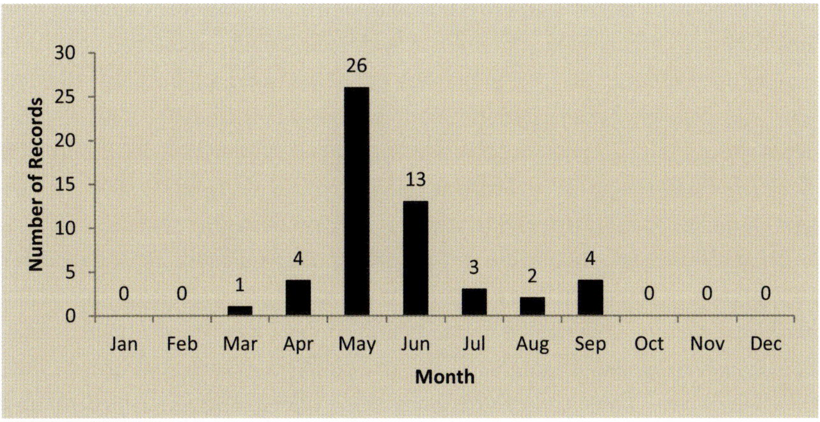

Collection/observation dates of 53 eastern musk turtles in Iowa.

nesting habitat in the form of uplands with sandy loam or silty loam soils adjacent to suitable aquatic habitat (Sutton 1994; Sutton and Christiansen 1999). Muskrat and beaver lodges and bank burrows may be used at times as shelters.

BEHAVIOR. When handled, some eastern musk turtles are shy and will pull their heads into their shells, although most may bite and scratch fiercely. The jaws are powerful and able to deliver a strong bite, with the turtle sometimes withdrawing its head into its shell as it holds on. They also release a foul-smelling fluid from glands located on the plastral bridge beneath the edge of the carapace—*odoratus* is Latin meaning "to have a strong odor." Chemicals in the fluid produce a strong unpleasant odor that has led to the use of the common names stinkpot and stinking Jim. Because the fluid evaporates quickly when exposed to air, the odor may be more effective at deterring terrestrial rather than aquatic predators (Eisner et al. 1977).

In northern states such as Michigan, Ohio, and Pennsylvania, the turtle has been reported to be active for around 200 days per year (Ernst and Lovich 2009). The collection/observation dates of 53 Iowa eastern musk turtles suggest a similar activity period, with the species found in the state from March through September. The earliest record is on March 8 in Clayton County and the latest is on September 11 in Muscatine County, though it is likely that they remain active in Iowa into October in most years.

Eastern musk turtles are secretive and highly aquatic and rarely leave the water, which likely contributed to Reeve Bailey not finding the species in Iowa during his surveys in the 1940s. They are bimodal breathers capable of taking up oxygen and excreting carbon dioxide in both air and water. While submerged, they only require one-eighth of the oxygen they use out of the water, which allows them to remain underwater for long periods (Root 1949). Root concluded that the remainder of the oxygen consumed while underwater occurs by absorption through the skin. The respiratory studies of Stone and his colleagues (1992) concluded that the turtles' reduced plastron with large areas of skin between the scutes creates a larger surface area for cutaneous oxygen absorption and consequently a greater ability for and a higher reliance on aquatic respiration. Although musk turtles may be able to remain submerged for several hours, most dives last less than 20 minutes. Poorly oxygenated water can have dire consequences for the turtles, with those in anoxic—containing no oxygen—water dying within 5 days (Ernst and Lovich 2009).

Eastern musk turtles are largely crepuscular but may become more diurnal in spring and fall at northern latitudes such as Iowa. They typically spend much of the day buried in the mud and become active in the late evening and early morning. The highly aquatic habits of these turtles may be related to water conservation. Eastern musk turtles tend to lose more water due to evaporation than do more terrestrial species (Ernst 1986). Even when their ponds dry up, these turtles may not seek new water, preferring instead to burrow into the mud (Gibbons et al. 1983). Nevertheless, some males have been documented moving up to 0.4 mile (0.7 km) overland after their ponds have dried up (Ernst 1986).

No data exist on Iowa home ranges, but in Pennsylvania home range was shown to vary between the sexes and habitats, with males using 4.3 acres (1.75 ha) and females using 2.3 acres (0.94 ha) throughout the year (Ernst 1986). The homing ability and tendency for eastern musk turtles to remain in or return to the area of their birth are well documented. In Michigan, homing ability was found in 75 percent of the eastern musk turtles studied with displaced individuals moving an average of 400 feet (122 m) per day as they returned to their home area (Williams 1952), and five eastern musk turtles displaced after an oil spill in the Kalamazoo River exhibited homing behavior, with two—one male and one female—traveling approximately 4.2 miles (6.7 km) and 3.5 miles (5.6 km), respectively, to return to their original point of capture (Otten and VanDeWalle 2014).

Both adult and juvenile eastern musk turtles bask to regulate body temperature, although not as frequently as some turtle species. Most often basking occurs in shallow water with only the top of the carapace exposed. Individuals will also leave the water and bask on exposed mud banks near the water's edge or on floating vegetation. Less commonly, they will bask above the water on a branch or other vegetation. Whether basking on the bank or on a branch, the turtle will quickly slip or drop into the water when disturbed. Eastern musk turtles have been found to be active from 50 to 93° F (10–340° C) with a preferred body temperature of 75.4° F (24.10° C; Ernst and Lovich 2009), using behavior to thermoregulate by moving from warmer to cooler water or vice versa to maintain an optimum body temperature. Even in habitats where the water temperature is below the preferred range, stinkpots have shown the ability to maintain their body temperature within their preferred range most of the time (Picard et al. 2011). Picard and his colleagues (2011) demonstrated that aquatic habitats with the highest thermal quality include those with surface cover, such as lily pads and other floating plants, followed by shallow-water areas, and as expected the turtles preferred these areas.

Eastern musk turtles overwinter underwater buried in the mud, under streambanks, or in muskrat lodges. During Iowa's cold winters, the turtles likely require places with well-oxygenated water that will not freeze to the bottom. As fall advances and the water gets colder, the turtles begin looking for a place to overwinter, beginning to burrow when the water temperature drops below 50° F (10° C; Ernst and Lovich 2009). They may sometimes congregate in large numbers in good overwintering habitat. Thomas and Trautman (1937) found 450 buried in the mud within a small area in a canal in Ohio that was partially drained during the winter. All but 6 were dead, apparently due to exposure to the cold air after removal of the water.

THREATS. Eastern musk turtle eggs are eaten by a variety of mammals, including raccoons, striped and spotted skunks, red and gray foxes, and even muskrats, reptiles such as snapping turtles and snakes, and birds such as American crows and herons. Adults and juveniles are also prey for raccoons, along with largemouth bass, bullfrogs, snapping turtles, northern watersnakes, hawks, and even bald eagles (Ernst and Lovich 2009). Plummer and Goy (1997) reported a dead eastern musk turtle found with its head inside the closed shell of a living mussel—apparently an accident when the turtle put its head inside the open mussel and the mussel then closed and trapped it inside, causing the turtle to suffocate or drown.

Habitat loss likely poses the most serious threat in Iowa through the draining and filling in of the wetlands and backwaters that are used by musk turtles as both summer and overwintering habitat, along with the loss of nesting habitat resulting from agricultural use of or development in sandy areas adjacent to summer habitat. Sedimentation of wetlands and streams causing shallower water that can freeze to the bottom can also result in mortality because freeze tolerance is

Eastern musk turtle from Muscatine County. Photo by Elizabeth Owens/TNC.

low in hatchlings and likely in adults as well. Increased flooding of nesting habitat as a result of drainage practices, stream alteration, or climate change and an increase in mammalian mesopredators over time, such as raccoons and skunks that depredate nests, also threaten the long-term stability of eastern musk turtle populations in Iowa.

FOOD AND FEEDING. Eastern musk turtles are bottom feeders that search for food primarily by smell while slowly walking through soft bottom sediments and probing with their snouts. They are omnivores, with small individuals feeding on aquatic insects, algae, and carrion and larger turtles eating whatever they find, including earthworms, snails, slugs, mussels, fish eggs, tadpoles, and a variety of aquatic insects and aquatic plants; they also scavenge dead animals when the opportunity arises. Jim recalls anglers in southeast Iowa catching stinkpots on hooks baited with earthworms. Prey is either swallowed whole or held in the jaws and torn into bite-sized pieces with the front feet. In Missouri, the most frequently eaten items were plants, insects, and mollusks (Ford and Moll 2004). Ford and Moll (2004) suggested that males preferentially consume soft-bodied insects while females eat more snails, but most other studies have found no sexual differences. Eastern musk turtles may opportunistically shift their diet to abundant foods, such as invasive mollusks, when available (Patterson and Lindeman 2009; Morrison et al. 2019).

REPRODUCTION. We have used descriptions of eastern musk turtle reproduction by Tinkle (1961, United States), Mahmoud (1967, Oklahoma), Ernst (1986, Pennsylvania), and Ernst and Lovich (2009) in this section except as noted. Maturation varies geographically, with turtles in the southern United States maturing faster than those in the north. Northern females reach sexual maturity in 2 to 7 years at straight-line carapace lengths greater than 3.1 inches (8 cm), while males in the north reach sexual maturity in 3 to 4 years at straight-line carapace lengths greater than 2.5 inches (6.3 cm). The single female in the Drake University Research Collection containing eggs is from Muscatine County and measured 4.5 inches (11.5 cm) straight-line carapace length.

Mating is generally concentrated in the spring, likely from late April into June in Iowa, but fall mating may also occur, with females storing sperm over the winter until ovulation in the spring. Mating occurs in shallow water with males using scent to locate females. Courtship includes three phases—tactile, mounting and intromission, and biting and rubbing. During the tactile phase, the male approaches his intended mate from behind with his neck extended, smelling or touching the tail to confirm the sex. If he has found a female, the male extends his head toward her musk glands on the plastral bridge. If unreceptive, the female moves away; however, the male may chase, nudge, and bite her. If she is receptive, the male approaches her from behind or the side and uses all 4 feet to grasp the margins of her carapace as he mounts. The male flexes a rear knee and holds the female's tail between the 2 small patches of rough scales on the inside of his hind legs, maintaining this position by looping his tail so that the hard tip on the end of the tail contacts the female near her cloaca. As the vents come together, the male inserts his penis—intromission—followed by rubbing and biting.

Ovarian follicles grow rapidly in the spring, reaching ovulatory size sometime in May. A Muscatine County female in the Drake University Research Collection found on May 19 contained 3 oviductal eggs and 2 follicles that had not yet been ovulated, suggesting that ovulation in Iowa likely occurs in May. Likewise, LeClere (2013) reported a road-killed Muscatine County female found on May 29 that contained a single egg. Sometime in June or July, females lay their eggs, with several females occasionally laying at the same site. Nests may be under logs or other objects, in rotting vegetation, or in a flask-shaped nest approximately 4 inches (10 cm) deep that the female digs with her back legs. Nests are located near the water and may be in the shade or in full sun, which can affect the sex of the hatchlings. Eastern musk turtles have temperature-dependent sex determination—eggs incubated at 82.4° F (28° C) or higher produce nearly all females, while those incubated at cooler temperatures produce both sexes. Eggs incubated at 77° F (25° C) produce up to 80 percent males.

Clutch size is proportional to female size, with larger females producing more eggs than smaller females. Across the range of the species, clutch size has been reported at 1 to 13 eggs, though most clutches average 2 to 4 eggs. The eggs are white, elliptical, 1 inch (2.6 cm) long on average, and hard-shelled. Clutch size seems to increase with an increase in latitude and a decrease in annual mean temperature, with more northerly populations having larger clutches on average. Clutches of 5 to 7 eggs (mean = 5.8; Illinois, Tucker et al. 2001) and 2 to 4 eggs (mean = 3.2; Pennsylvania, Ernst 1986) have been reported for northern populations. Iowa clutch size observations consist only of Dodge (1956a), who reported a Muscatine County female that laid 6 eggs and the Muscatine County female in the Drake University Research Collection with 3 oviductal eggs and 2 ovarian follicles below ovulatory size, suggesting a clutch size of 5 eggs. South of Iowa, females may lay multiple clutches per year, which seems unlikely in Iowa, where they probably lay only a single clutch due to the shorter activity season at the northwest edge of the species' range.

Eggs hatch in August and September after an approximately 65- to 86-day incubation period, with hatchlings ranging from 0.7 to 1 inch (1.7–2.6 cm) straight-line carapace length. Hatchlings in Iowa likely emerge from the nest in the fall, because overwintering in the nest is unlikely at northern latitudes since hatchlings are susceptible to death at low temperatures and also dehydrate readily. Dodge (1956a) reported hatching on September 14 and 15 with hatchlings ranging from 0.9 to 0.94 inch (2.3–2.4 cm) straight-line carapace length and 0.7 to 0.75 inch (1.7–1.9 cm) straight-line carapace width. Three eggs incubated in Jim's lab at Drake University hatched the first week of September, with hatchlings ranging from 0.96 to 0.98 inch (2.45–2.5 cm) straight-line carapace length and 0.7 to 0.75 inch (1.7–1.9 cm) straight-line carapace width. Noble and Breslau (1938) found that hatchling dispersal from nests was primarily oriented toward the light of the open horizon, with humidity a secondary stimulus.

LONGEVITY. A wild-caught adult survived in captivity an additional 54 years and 9 months. Recaptured wild individuals have been estimated at 27 to 28 years (Ernst 1986).

FAMILY TRIONYCHIDAE

SOFTSHELL TURTLES

The family Trionychidae arose in the early Cretaceous—145 to 100 million years before the present—and includes thirteen genera and thirty-three species found in North America, Africa, Asia, and the East Indies; the group diversified independently of other turtle lineages. Only one genus, *Apalone*, is found in North America; two species, the eastern spiny softshell and the midland smooth softshell, occur in Iowa.

Members of the family are a highly derived group—"derived" refers to traits that exist in a present organism but did not exist in that organism's ancestors—with a rounded and flattened carapace and a greatly reduced plastron, both of which lack horny scutes and are instead covered with leathery skin. The neck is long and retractable, the limbs are paddle-like with broadly webbed feet containing three claws, and the snout is long and shaped like pair of needle-nose pliers and can inflict a painful bite. The long tubular snout serves as a snorkel to allow the turtle to breathe while submerged. Trionychids are highly aquatic and strong fast swimmers. Their meat is highly sought after in many parts of the world, which has led to the decline of softshells in places where their harvest is not regulated.

MIDLAND SMOOTH SOFTSHELL

Apalone mutica mutica (Lesueur) 1827

Midland smooth softshell from Carroll County. Photo by Don Becker.

IOWA STATUS. Species of greatest conservation need. Harvesting legal with proper licenses.

HISTORICAL SUMMARY. The midland smooth softshell has also been called the unarmed trionyx, leatherback, and pancake turtle. Agassiz (1857) mentions receiving smooth softshells from Iowa and eggs from Burlington in Des Moines County. The first confirmed museum specimen in the state appears to be in 1883 near Davenport in Scott County (National Museum of Natural History–Smithsonian Institution). The softshell turtle was mentioned in the *History of Fremont County,*

Iowa (Anonymous 1881) as occurring in the county, although which species is unclear because the scientific name provided is an older name for the Florida softshell (*Apalone ferox*), a species not found in Iowa. Somes (1911) listed the species as present in Iowa, and Moorman and Carlander (1951) stated that they were "unpopular in the pond but welcome at the dinner table!"

Two subspecies of the smooth softshell, *Apalone mutica,* are currently recognized, only one of which, the midland smooth softshell, *A. m. mutica,* has a range that includes Iowa.

DESCRIPTION. Midland smooth softshells have a flat round leathery carapace that is keelless and a reduced hingeless plastron. Conant (1975) described them as "animated pancakes" in reference to their flat disc-like form. The carapace lacks a sandpapery feel, spines, or raised knobs; although the leading edge may have evenly spaced furrows, there are no spines or tubercles—*Apalone* is from the Greek *apalos* meaning "soft, tender"; *mutica* is Latin meaning "curtailed, unarmed" in reference to the lack of tubercles on the front edge of the carapace. The carapace is gray, tan, or olive-brown with a pattern of darker dots, dashes, or blotches, and there may be a lighter band surrounding the darker outside edge. The plastron is bright white or gray. Because of the relatively flexible shell, the vertebral column and ribs can often be seen under the skin of the carapace, and the underlying bones are also visible beneath the skin of the plastron. The upper side of the head, neck, and limbs is olive or gray and white or cream below, and a dark-bordered light stripe extends through the eye onto the neck. The mostly unpatterned legs end in webbed feet, with the webbing extending up the hind limbs. The elongated tubular snout ends somewhat diagonally with no horizontal ridge within the nostrils. Hatchlings are similar to adults with a nearly circular olive carapace marked with faint dots and dashes and a light-colored outer edge.

Midland smooth softshells exhibit pronounced sexual dimorphism, with males having a brown or grayish tan carapace with scattered small dark dots and dashes, longer foreclaws than females, and long thick tails with the vent located near the tip. Females are larger than males with a mottled gray, brown, or olive carapace, hind claws that are longer than those of males, and short tails with the vent located beneath the carapace.

SIZE. Female midland smooth softshells generally range from 6.5 to 14 inches (16.5–35.6 cm) straight-line carapace length; males generally range from 4.5 to 10.5 inches (11.5–26.7 cm) straight-line carapace length. The largest specimen in the Drake University Research Collection is a female from Louisa County measuring 10.4 inches (26.4 cm) straight-line carapace length, though larger individuals are

Plastron of a midland smooth softshell from Linn County. Photo by Don Becker.

certainly found in Iowa. The maximum straight-line carapace length reported for the species is 14 inches (35.6 cm).

SIMILAR SPECIES. Eastern spiny softshells are the only other Iowa turtle with a soft leathery carapace. In contrast to the midland smooth softshell, the carapace of the eastern spiny softshell has a sandpapery feel, spines or tubercles along the leading edge, and darker and more circular markings. The skin of the upper side of the head, neck, and limbs of the eastern spiny softshell is olive to tan and patterned with dark spots and streaks; the throat is grayish white with small dark gray markings. In addition, a horizontal ridge is visible within each nostril.

DISTRIBUTION. *A. m. mutica* is found principally in the Missouri, Mississippi, and Ohio Rivers and their larger tributaries, along with some rivers in eastern Texas that drain to the Gulf of Mexico. West of the Mississippi River, the range extends into portions of North and South Dakota, Nebraska, Kansas, Oklahoma, and central Texas. East of the Mississippi River, the turtle is found in large Mississippi River tributaries in Wisconsin and Illinois as well as in the Ohio River and its tributaries into southeast Ohio and south into portions of Kentucky and Tennessee. Scattered populations are also found in Mississippi and Alabama.

In Iowa, midland smooth softshells are found in scattered populations in the larger rivers of the state, including the Mississippi, Cedar, Iowa, Des Moines, Raccoon, Skunk, Boone, and Missouri Rivers. In the mid-1940s, the species had been documented in 8 counties (9 Mississippi River records and 1 each in the Boone and Raccoon Rivers); at present, records exist for 35 of Iowa's 99 counties. In the early 1980s, based on sampling conducted between 1970 and 1980, Jim Christiansen (1981) suggested that midland smooth softshell populations in Iowa were stable. In the 1990s, however, Terry and Jim (1996) found that midland smooth softshells were one of the turtle species most affected by stream modification in the state and that they had been eliminated or severely reduced in number in the most heavily modified rivers.

Midland smooth softshell distribution in Iowa.

Blue = pre-1960
Yellow = post-1960
Green = both pre- and post-1960

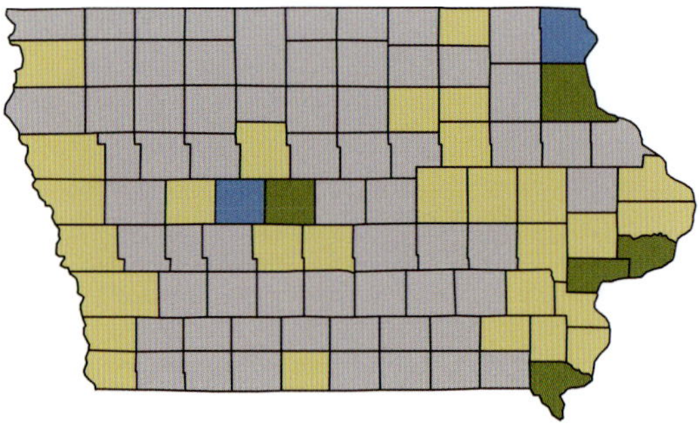

HABITAT. Midland smooth softshells are riverine turtles that most commonly inhabit the open water areas of midsize to large rivers with moderate to fast currents. During flooding, they may inhabit floodplain impoundments that are temporarily connected to the main channel. In Iowa, they avoid small streams and tributaries and are associated with larger rivers containing extensive sandbars. Tom Williams and Jim (1981) captured midland smooth softshells in Iowa only in open water areas near sandy beaches, whereas eastern spiny softshells were caught in areas with abundant submerged brush, trees, and other debris. Similarly, Anderson et al. (2002) found that midland smooth softshells comprised 80 percent of the turtles captured in the open water habitat of a reach of the upper Mississippi River across from Keokuk, Iowa. Barko and Briggler (2006) found that midland smooth softshell abundance in the Mississippi River was greatest in deep waters with faster velocity compared to eastern spiny softshells, which were more abundant in waters with higher visibility and slower velocity.

Habitat use varies with sex and age. In Kansas, males were captured in all habitats but were most abundant in shallow water near sandbars (Plummer 1977a). In

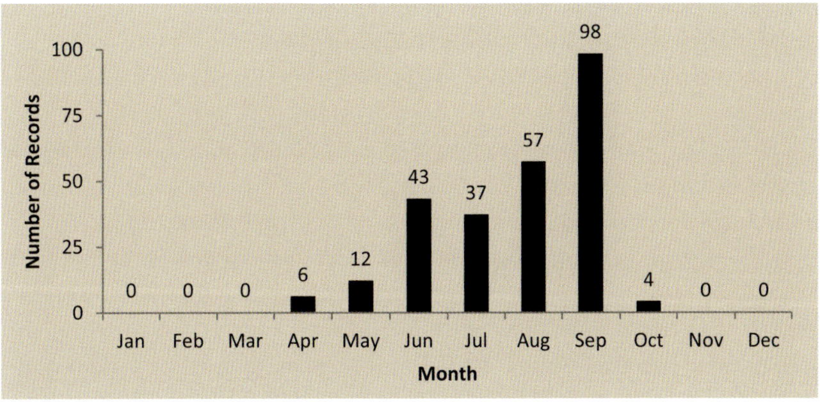

Collection/observation dates of 257 midland smooth softshells in Iowa.

contrast, females preferred deeper, more open water except during the nesting season, when they used shallow water near sandbars. Hatchlings preferred the warmer, quieter water of small shallow puddles on sandbars or the shallow water on the downstream end of a sandbar, where they often burrowed into the sand at the water's edge.

Sand is an important component of the habitat of the midland smooth softshell. The turtles are most often found in waterways with soft sandy or gravelly bottoms. Further, they bask almost entirely on sandy or muddy shorelines or in shallow water associated with such shorelines, nest on clean sandbars, and burrow in sand to avoid predation and strong water currents or to thermoregulate. Plummer et al. (2008) concluded that both juveniles and adults grow faster and have healthier shells when they have ready access to a sandy substrate in which to burrow.

BEHAVIOR. Midland smooth softshells are the more aquatic of Iowa's two softshell species, rarely leaving the water except to nest or bask. As such, they are seldom found far from the water, and although not well adapted to moving on land, if startled while basking they quickly make a dash for the water. If handled, they can rapidly extend their long necks and deliver a forceful bite, though unlike the eastern spiny softshell they are more likely to withdraw into their shells rather than bite. Nevertheless, docile behavior is not a given, and they may also use their sharp claws for defense.

The collection/observation dates of 257 Iowa midland smooth softshells show that the species has been found from April to October in Iowa, though they are most active from May through September, similar to Kansas where they become active in late March or April at water temperatures between 56.3 and 59° F (13.5–15° C; Plummer 1977a). The earliest Iowa record is one found on April 3 in Boone County.

The turtle is almost entirely diurnal, spending much of its day searching for food or basking near the water on the sandy or muddy banks of a river, often with its neck fully outstretched. Tom Williams and Jim (1981) observed that midland smooth softshells in Iowa basked solely on sandy or muddy beaches or in shallow water associated with such beaches, and individuals avoided basking on objects such as logs, unlike eastern spiny softshells. Similarly, Lindeman (2001) also reported that smooth softshells basked more often on shores than did spiny softshells. Plummer (1977a) found that males more often basked on sandbars, whereas females more frequently basked on steeper banks near deep water.

When not basking or hunting, the turtle rests in shallow water, where it burrows into loose sand on the bottom of the river by first pushing its head and front legs into the sand and then shuffling its body to stir up sand and other debris. The stirred-up material settles on top of the carapace, concealing the turtle. Only the head is then visible, and the shallow water allows the turtle to extend its snorkel-like snout to the surface to breathe. The large surface area of the skin, including the skin covering the shell, is very permeable to gases and water, allowing oxygen and carbon dioxide to pass across the skin into and out of the water and enabling the turtle to remain submerged for long periods. When it is out of the water, however, the extensive amount of skin promotes water loss, limiting the amount of time a softshell can stay on land.

Throughout the active season, midland smooth softshells occupy linear home ranges within the rivers in which they live. Overall home range length varies from approximately 1.2 to 1.9 miles (2–3 km); however, differences are seen between males and females and between subadult and adult females. Adult females use the largest home ranges, followed by subadult females and then males (Plummer and Shirer 1975). In addition, the movement per day within a home range differs between the sexes—females move significantly more than males and tend to occupy both sides of the river, while males tend to stay on one side. Radiotelemetry of a single midland smooth softshell in Iowa suggested that it was considerably more active than an eastern spiny softshell that was tracked concurrently (Williams and Christiansen 1981). In addition, midland smooth softshell activity was concentrated in the water column closer to the water surface, in contrast to the eastern spiny softshell, which appeared to spend more time on or near the bottom in cooler water.

Juvenile midland smooth softshell from Linn County partially buried in the sand. Photo by Mike Pingleton.

As water temperatures begin to drop in late fall, the turtles become less active and likely begin overwintering sometime in late September or early October in Iowa. In Kansas, activity was last observed in October at water temperatures between 66.2 and 68° F (19–20° C; Plummer 1977a). The latest Iowa records are several hatchlings found in Linn County on October 9. Midland smooth softshells overwinter buried in the sediment on the river bottom in areas where the current is sufficiently strong to provide adequate oxygen. The water must also be deep enough to prevent freezing to the bottom, which would also prevent gas exchange. While the turtles are submerged during the winter, oxygen is extracted from the water and carbon dioxide from the blood through a combination of cutaneous and buccopharyngeal respiration.

THREATS. American crows, eastern moles, raccoons, skunks, opossums, and domestic dogs all eat midland smooth softshell eggs and can be significant nest predators. Goldsmith (1944) described the predation of midland smooth softshell nests along the Mississippi River near Dubuque by what he believed to be one large raccoon that was "feasting" nightly on the eggs. He said, "It was obvious by the many heavy imprints of the left foot in the sand on the left side of the excavated nests and the almost trackless right side that the raccoon had used his right 'hand' to excavate and to remove the eggs from the nest. The shells were usually placed in one pile four or five inches from the hole on the right side. These eggs seemed to be equally appetizing whether they contained embryos or turtles about ready to hatch." Juveniles may be consumed by various predators, including large fish, watersnakes, snapping turtles, wading birds, and a variety of mammals. Thirty-six softshell turtle shells, including midland smooth softshells, were found in a bald eagle nest in southeast Wisconsin (Watermolen 2004). Nevertheless, due to their highly aquatic lifestyle and swimming abilities, adults are likely not commonly depredated.

Humans pose a significant threat to midland smooth softshells through habitat loss resulting from the modification of the riverine habitats in which they live, overharvesting, and climate change. In the 1990s, Terry (1993) studied the effect of stream modification on riverine turtles in Iowa. The work showed that activities such as channelization, riprapping of streambanks, and impounding of rivers result in a reduction or loss of diversity of both the types of habitats and the turtle species present in riverine systems. As the habitats become more homogeneous and less complex, rare species or those with specialized habitat requirements are lost. Midland smooth softshells have specific habitat requirements that are often no longer present after stream modification, and they are particularly affected by it. Clean sandy beaches are essential nesting areas, and channelization and the construction of levees and dikes confine floodwaters to the main channel of the river, resulting in scouring effects that remove the loose sand that would otherwise be deposited as sandy shores. The consequence is a reduction in the number of suitable nest sites. In addition, the suitable sites that remain are often covered with water during the nesting period and, as the high water recedes, silt is deposited that covers the sandbars and creates unsuitable nesting conditions. Further, because midland smooth softshells hunt in the water column, the silt-laden water may impair their ability to locate prey.

Until recently, commercial harvesting was likely a major threat in Iowa. Based on Iowa Department of Natural Resources records, between 1987 and 2017 commercial turtle trappers harvested an average of 26,281 pounds (11,921 kg) of softshell turtles per year in the state. Because harvesters were not required to distinguish between midland smooth and eastern spiny softshells, it is not possible to know exactly how many midland smooth softshells were taken. However, the reported average does illustrate the magnitude of the softshell catch. Changes to Iowa's turtle harvest regulations in 2017 made a significant difference. In 2021, commercial turtle trappers harvested just 16 pounds (7 kg) of softshell turtles.

Recreational harvesters are not required to report their catch; therefore, the number of softshell turtles taken by recreational harvesters is unknown, though under current regulations recreational and commercial collection of softshell turtles is closed from May 15 to July 15. Outside of these dates, the rules limit the daily catch to one eastern spiny softshell or one midland smooth softshell for both recreational and commercial harvesters.

Climate change is an emerging threat that could affect populations in years to come. Increased rainfall and more severe rain events may increase the frequency and duration of river flooding, which could reduce the availability of suitable nest sites.

FOOD AND FEEDING. The cryptic coloration and bottom-dwelling behavior of the midland smooth softshell are ideal adaptations for this sit-and-wait ambush predator. As prey passes by, the turtle quickly extends its neck and grasps it. Small animals may simply be sucked in when the turtle rapidly expands its mouth-pharynx region, creating a low-pressure area that pulls the food into its mouth in a pharyngeal gulp. The turtles may also cruise bottom debris or sandbars in search of food. Jim recalls observing midland smooth softshells poking their long necks into the sand or into holes in the mud in search of prey. If the prey flees, these powerful swimmers chase and capture it. They may at times forage on land as well.

Midland smooth softshell from Linn County. Photo by Mike Pingleton.

These turtles are primarily insectivores, consuming a large variety of aquatic insects along with other aquatic invertebrates such as worms, mollusks, and snails. Fish, amphibians, young birds, mammals, algae and other plants, fruit, and carrion have also been reported in the diet (Ernst and Lovich 2009). Based on stomach content analysis, Williams and Christiansen (1981) concluded that the diet of Iowa midland smooth softshells was about 75 percent insects, including many species that occupy the water column. They went on to speculate that the turtle mainly forages in large areas of open water containing a current. Approximately 20 percent of the stomachs examined contained small fish, which were thought to have been eaten alive. Only 8 percent of the stomachs contained carrion, and 12 percent contained plant material, suggesting accidental ingestion. Kansas studies showed differences in diet between males and females (Plummer and Farrar 1981). Males have a more diverse diet consisting of more terrestrial prey, while females consumed more aquatic prey items. The differences in diet likely reflect differences in habitat use—females spend more time foraging in deeper water; males spend more time foraging in shallow water near aquatic-terrestrial interfaces.

REPRODUCTION. Males reach sexual maturity in their fourth year at around 3.1 inches (8 cm) straight-line plastron length; females mature during their ninth year at around 5.5 inches (14 cm) straight-line plastron length (Plummer 1977c). Ovulation occurs in the spring, likely May, after emergence from overwintering.

Mating takes place in the water in April or May, shortly after spring emergence, and possibly again in the fall. As described by Plummer (1977b), males actively search for females in the water and at basking sites, investigating the sex of prospective mates by probing under the edges of carapaces. When males probe other males, the investigation may either be ignored or the basking male may confront the probing male with an open-mouthed aggressive display. A female, if unreceptive, will attack the would-be suitor, sometimes inflicting semicircular wounds on the rear edge of the male's carapace as she chases and bites him. If receptive, the female becomes passive as the male mounts her. The male approaches from the rear, crawls onto the female's back, and positions his tail beneath hers to align the vents. Because the male does not grasp the female with his forelimbs, he is forced to swim in place atop her to maintain his position—possibly up to 20 minutes. Copulation seems to occur in deeper water and lasts at least 30 minutes.

Sometime in late June or early July, females lay their eggs, often with many females laying at the same site. In Dubuque County, Goldsmith (1944) noted, "Often times the nests are so close together, one spade full of sand might unearth three or four nests, thus giving the popular impression that one turtle lays almost a hundred eggs." Nests are constructed on clean gently sloping sandbars or sandy shores free of weeds, debris, and sediment and offering an unobstructed view of the water and up and down the shoreline. Suitable nesting sites have sand damp enough to cling together but without any clay or organic matter that can encase the eggs in packed sand and, thereby, prevent emergence of the hatchlings (Muller 1921). Nesting occurs in the morning or early afternoon, at which time females dig nests 6 to 10 inches (15.3–25.4 cm) deep with their hind feet within 100 feet (30.5 m) of the water's edge and above the normal river level. At Mississippi River locations adjacent to Iowa, Goldsmith (1944) noted that nests were 6 to 25 feet (1.8–7.6 m) from the water, and Muller (1921) found nests 10 to 60 feet (3–18.3 m) inland. Each female may dig several test holes before finding a site she likes. As eggs are laid, the female uses her hind feet to position them in the nest. Once the eggs have been laid, she uses her hind feet to fill in the nest.

Females lay from 4 to 33 ovoid to spherical eggs 0.9 inch (2.3 cm) in diameter with thick brittle white shells. Clutch size is proportional to female size, with larger females producing more eggs than smaller females. In the Midwest, clutches of 3 to 26 eggs (Kansas, Plummer 1977c), 18 to 31 eggs (Illinois, Cahn

Hatchling midland smooth softshell from southeast Iowa showing the dot and dash pattern on the carapace. Photo by Ryan D. Rasmussen.

1937), 6 to 26 eggs (Illinois, Phillips et al. 1999), 15 to 25 eggs (Wisconsin, Vogt 1981b), and 10 to 11 eggs (Wisconsin, Paisley and Hay 2022) have been reported. Muller (1921) reported 4 to 33 eggs for nests on a Mississippi River island on the Illinois side approximately one mile north of Fairport, Iowa. In Iowa, Goldsmith (1944) reported 10 to 31 eggs for 11 clutches (average = 18.7) in Dubuque County, and LeClere (2013) described 15 clutches in Muscatine County containing 11 to 24 eggs (average = 18).

Eggs hatch in August and September after approximately 44 to 82 days, depending on incubation temperature. The greatest hatching success occurs between 80.6 and 91.4° F (27–33° C; Janzen 1993; Plummer et al. 1994). Midland smooth softshells appear to have genetic sex determination. While incubation temperature does not affect the sex of the hatchlings, within nonlethal limits embryonic growth, body size, and both running and swimming speed were positively correlated to incubation temperature (Janzen 1993). The hatchlings, which range from 1.2 to 1.5 inches (3–3.7 cm) straight-line carapace length, lack a caruncle—an egg tooth, a temporary keratinized projection on the snout used to cut the eggshell at the time of hatching—and must use their forelimbs to break through the eggshell. Hatchlings emerge from the nest in fall, leaving the nest within the first 3 hours after sunset and all emerging within 2 hours (Anderson 1958). Muller (1921) believed that the hatchlings exhibited "a marked geotropism, always going downhill, and are photokinetic." He noted that nests are usually situated where an open sky and a sloping beach are presented in the direction of the water and went on to speculate that it is probable that these two factors guide the hatchlings to the river. Anderson (1958), however, found no evidence of phototropism.

LONGEVITY. Softshell turtles do not establish growth rings as do hard-shelled turtles, so longevity in the wild is uncertain. Some *A. mutica* have lived 11 years in captivity, and survival of over 20 years in the wild is suggested (Ernst and Lovich 2009).

EASTERN SPINY SOFTSHELL

Apalone spinifera spinifera (Lesueur) 1827

Eastern spiny softshell from Marion County.

IOWA STATUS. Species of greatest conservation need. Harvesting legal with proper licenses.

HISTORICAL SUMMARY. The eastern spiny softshell has also been called the leatherback and the pancake turtle. Agassiz (1857) mentions receiving spiny softshells from Iowa, though the first confirmed museum specimens in the state appear to be in 1878 near Webster City in Hamilton County and 1883 near Davenport in Scott County (National Museum of Natural History–Smithsonian Institution). The softshell turtle was mentioned in the *History of Fremont County, Iowa* (Anonymous 1881) as occurring in the county, although which species is unclear because the scientific name provided is an older name for the Florida softshell (*Apalone ferox*), a species not found in Iowa. Yarrow (1882) reported the Hamilton County specimen in the National Museum of Natural History, Osborn (1892) referenced an Iowa State Natural History Collection specimen believed to be from Ames in Story County, Somes (1911) stated that the species was found in Iowa but did not reference a specimen, and Moorman and Carlander (1951) stated that it was "unpopular in the pond but welcome at the dinner table!"

Five subspecies of the spiny softshell, *Apalone spinifera*, are currently recognized, only one of which, the eastern spiny softshell, *A. s. spinifera*, has a range that includes Iowa.

DESCRIPTION. Eastern spiny softshells have a carapace that is flat, round, keelless, and leathery and a reduced hingeless plastron. Conant (1975) described them as "animated pancakes" in reference to their flat disc-like form. The carapace has a sandpapery feel and conical spiny projections along the leading edge—*Apalone* is from the Greek *apalos* meaning "soft, tender"; *spina* is Latin meaning "thorn" or "spine"; and *-ifer* is Latin meaning "bearing" in reference to the spine-like projections along the front edge of the carapace. In general, the carapace is olive to tan with a pattern of black ocelli—circular markings—or dark blotches. A dark line surrounds the margin of the carapace; however, there is pronounced sexual dimorphism. Males have an olive-gray carapace with black ocelli and a distinct sandpapery feel and long thick tails with the vent located near the tip; females are larger than males with a dark olive or tan carapace with brown and gray mottling

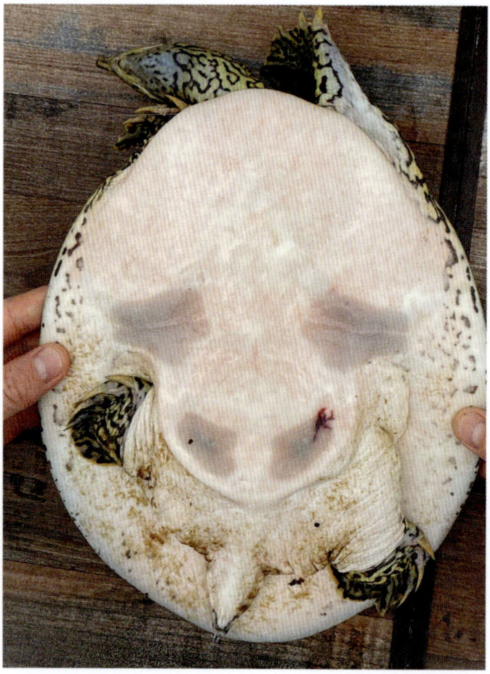

Plastron of an eastern spiny softshell from Linn County. Photo by Daniel F. Hughes.

that is smooth or has only a slight sandpapery feel and short tails that barely extend past the edge of the carapace, and the vent is located beneath the carapace.

On both sexes, the plastron is bright white or yellow, and the underlying bones may be visible beneath the skin. The upper side of the head, neck, and limbs is olive to tan and patterned with dark spots and streaks; the throat is grayish white with small dark gray markings. Two dark-bordered light stripes are present on each side of the head—one extending backward through the eye and the other backward from the angle of the jaw—and the lips are yellowish. The patterned legs end in webbed feet, with the webbing extending up the hind limbs. The shortened tubular snout has large nostrils with a horizontal ridge visible within each nostril. Hatchlings resemble adult males with circular markings on the carapace and a pronounced sandpapery feel, but they have poorly developed spines along the anterior edge.

SIZE. Female eastern spiny softshells generally range from 7 to 21.25 inches (18–54 cm) straight-line carapace length; males generally range from 5 to 12.2 inches (12.5–31 cm) straight-line carapace length. The largest specimen in the Drake University Research Collection is a female from Muscatine County measuring 12.5 inches (31.7 cm) straight-line carapace length, though larger individuals are undoubtedly found in the state. The maximum straight-line carapace length reported for the species is 21.25 inches (54 cm).

SIMILAR SPECIES. Midland smooth softshells are the only other Iowa turtle with a soft leathery carapace. In contrast to the eastern spiny softshell, the carapace of the midland smooth softshell lacks a sandpapery feel and spines or tubercles along the leading edge and has a pattern of darker dots, dashes, or blotches. The head, neck, and limbs of the midland smooth softshell are olive or gray on the upper side and white or cream below. In addition, it lacks a horizontal ridge within the nostrils.

Nostrils of an eastern spiny softshell from Muscatine County showing the horizontal ridges. Photo by Ryan D. Rasmussen.

DISTRIBUTION. The eastern spiny softshell occurs more or less continuously from central Minnesota, Wisconsin, Michigan, and New York southward to Oklahoma, Arkansas, northern Mississippi, and Alabama and from the Appalachian Mountains west to the middle of Nebraska, Kansas, and Oklahoma. It occurs in the west in the big rivers of New Mexico, Colorado, and southeast Wyoming with disjunct populations in the big rivers of western South Dakota and Montana. In the east, disjunct populations are found in southern Quebec, New Hampshire, and Maryland.

In Iowa, eastern spiny softshells are found statewide. In the mid-1940s, the species had been documented in 23 counties; at present, records exist for 74 of Iowa's 99 counties. In the early 1980s, based on sampling conducted between 1970 and 1980, Jim Christiansen (1981) suggested that eastern spiny softshell populations in Iowa were stable. In the 1990s, Terry and Jim (1996) found that eastern spiny softshells were one of the turtle species least affected by stream modification, and they are still commonly found across the state where suitable habitat exists.

Eastern spiny softshell distribution in Iowa.

Blue = pre-1960
Yellow = post-1960
Green = both pre- and post-1960

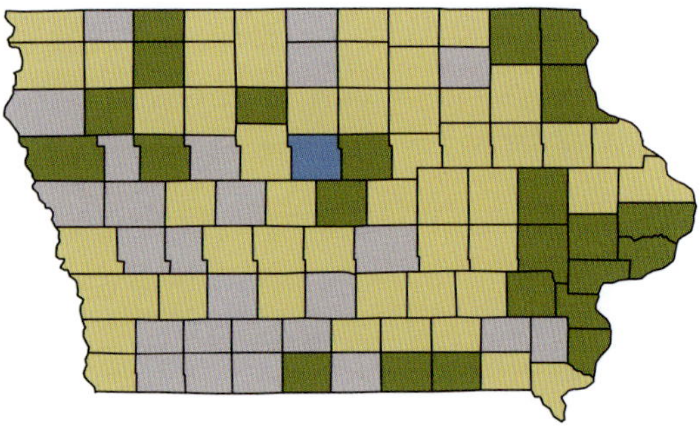

HABITAT. Eastern spiny softshells are primarily a riverine species, but they may also be found in smaller streams, marshes, flooded road ditches, lakes, ponds, and impoundments in Iowa. Even in larger rivers, they tend to prefer slow-moving backwater areas. Barko and Briggler (2006) found that eastern spiny softshell abundance in the Mississippi River was greatest in waters with higher visibility and slower velocity compared to midland smooth softshells, which prefer deep water with faster velocity. In a study comparing habitat use of eastern spiny softshells and midland smooth softshells in Iowa, Tom Williams and Jim (1981) found that the eastern spiny softshell used a broader variety of habitats, frequenting not only large rivers but small streams as well. In addition, they were caught in areas with abundant submerged brush, trees, and other debris,

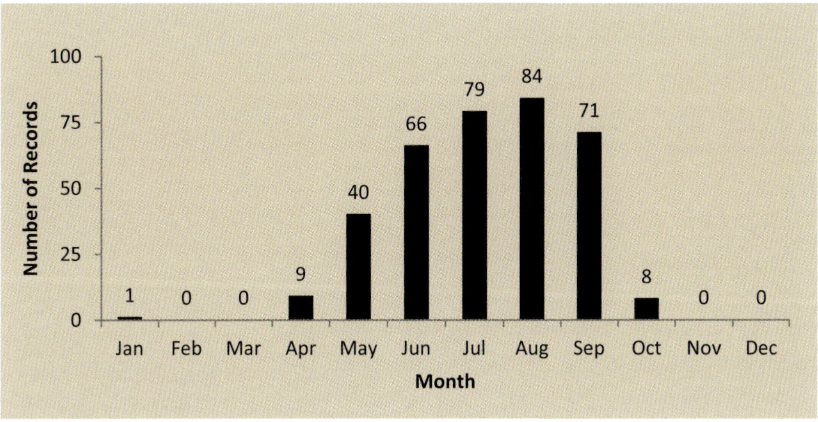

Collection/observation dates of 358 eastern spiny softshells in Iowa.

whereas midland smooth softshells were caught only in open water areas near sandy beaches. Habitat use seems to vary between the sexes, with females preferring open water more than males and males spending more time in riffles than females do (Williams and Christiansen 1981; Plummer et al. 1997). Waterbodies with sandbars and mudflats for basking, at least some aquatic vegetation and submerged trees, and a soft sand, silt, or mud bottom that allows for easy burying are preferred.

BEHAVIOR. These highly aquatic turtles are rarely found far from the water. Although not well adapted to moving on land, if startled while basking they quickly sprint for the water. When Neil has inadvertently captured them while seining fish, the turtles begin a rapid swim-crawl-climb that propels them around the net much like a hamster in an exercise wheel. If handled they can be quite aggressive, rapidly extending their long necks backward about half the length of their shells and delivering a painful bite similar to being grabbed by a pair of needle-nose pliers. Terry had a coworker who was once bitten on the forearm by a large eastern spiny softshell that left a gash of approximately 1 inch (2.6 cm) requiring several stitches to close up. In addition to biting, they scratch fiercely with their sharp claws in an attempt to get away.

The collection/observation dates of 358 Iowa eastern spiny softshells show that the species has primarily been found from April to October in Iowa, though they are most active from May through September. Nevertheless, the earliest Iowa record is a softshell found by Everett Speaker—one of the authors of *Iowa Fish and Fishing* (1987)—on January 17 in a Mississippi River slough in Scott County; it is not known whether the turtle was active or hibernating at the time it was found.

Eastern spiny softshells are almost entirely diurnal, spending much of the day searching for food or basking near the water on a sandy or muddy shoreline, rock, log, or floating debris or floating on the surface in shallow water. In Iowa, Tom Williams and Jim (1981) observed them basking on partially submerged trees, rocks, floating logs, or other floating vegetation, though these sites were never used when sandy beaches or sandbars were available. When not basking or hunting, the turtle rests in shallow water, where it buries itself in the bottom material by first pushing its head into the sand or mud, then retracting it while powerfully pulling its forelimbs and pushing its hind limbs to crawl beneath the bottom material (Graham and Graham 1991). After moving forward several inches, the turtle employs a rapid "waggle" or "shimmy" of its rear end so that the bottom material settles smoothly over its carapace. When the turtle is buried, only its head may be visible, and the shallow water allows it to extend its snorkel-like snout to the surface to breathe. Neil recalls paddling shallow clear streams in Iowa and seeing what appeared to be a round piece of stream bottom suddenly swim away, easily outdistancing his canoe. At night, the turtle sleeps buried in the bottom material or among the branches of submerged trees.

The large surface area of the turtle's skin, including the skin covering the shell, is very permeable to gases and water, allowing oxygen and carbon dioxide to pass across the skin into and out of the water, enabling the turtle to remain submerged for long periods. Eastern spiny softshells have a remarkable ability to extract oxygen from the water, with up to 38 percent of their oxygen exchange and 85 percent of their carbon dioxide exchange occurring through their skin while they are submerged (Ernst and Lovich 2009). When they are out of the water, however, the extensive amount of skin promotes water loss at an extremely high rate compared to other reptiles (Plummer et al. 2005), limiting the amount of time that a softshell can stay on land. The high evaporative water loss also causes softshells to cool faster than they heat in air, thereby promoting loss of body heat, whereas in the water they heat faster than they cool (Smith et al. 1981). Based on limited temperature data, Tom Williams and Jim (1981) reported a range of water temperatures for active Iowa eastern spiny softshells mainly between 77 and 89.6° F (25–32° C). Body temperature of Arkansas eastern spiny softshells varied between 67.8 and 84.7° F (19.9–29.3° C) based on activity—buried, swimming, or basking (Plummer et al. 2005). In lab experiments, Iowa juvenile eastern spiny softshells preferred the warmest temperatures offered (Feltz and Tamplin 2007).

Eastern spiny softshells are powerful swimmers that use their forelimbs to pull them through the water, and movement is largely accomplished by swimming. While they do use a home range, they tend to move less than midland smooth softshells. Males and females in a small Arkansas stream moved 85 percent of the day, with males averaging 463 feet (141 m) and females averaging 400 feet (122 m) per day (Plummer et al. 1997). Radiotelemetry of a single eastern spiny softshell in Iowa suggested considerably less activity than a midland smooth softshell that was tracked concurrently (Williams and Christiansen 1981). In addition, the eastern spiny softshell appeared to spend more time on or near the bottom in cooler water, in contrast to the midland smooth softshell, which spent more time in the water column closer to the surface. Joshua Otten and Terry (2014) provided the first report of presumed homing behavior in the eastern spiny softshell following an oil spill in the Kalamazoo River in Michigan. Three individuals traveled from 15.7 to 28.2 miles (25.25–45.4 km) from their release points back in the direction of their initial capture point. These distances represented the furthest reported movements of eastern spiny softshells, with 28.2 miles (45.4 km) being one of the longest homing movements reported for a freshwater turtle in the United States.

As water temperatures begin to drop in late fall, the turtles begin moving to their overwintering locations, which may be some distance from their summer activity areas. In Iowa, they likely begin overwintering sometime in late September or early October. The latest Iowa record is a hatchling found in Linn County on October 9. Eastern spiny softshells overwinter underwater buried approximately 2 to 4 inches (5–10 cm) in the sediment. The turtles cannot withstand freezing and are the most intolerant of anoxia—lack of oxygen—of all North American turtles (Reese et al. 2003). Therefore, to survive overwintering, they require water with sufficient current or depth to prevent freezing and a high oxygen content. Jim and John Bickham (1989) documented winterkill of eastern spiny softshells at Big Sand Mound Nature Preserve in Muscatine and Louisa Counties during an unusually cold winter that was preceded by wetland drying, resulting in low winter water levels. While the turtles are submerged during the winter, oxygen is extracted from the water and carbon dioxide from the blood through a combination of cutaneous and buccopharyngeal respiration.

THREATS. Eastern moles, raccoons, skunks, opossums, foxes, and domestic dogs all eat eastern spiny softshell eggs and can be significant nest predators. Juveniles are consumed by various predators, including large fish, watersnakes, snapping turtles, wading birds, muskrats, and other mammals. Thirty-six softshell turtle shells, including eastern spiny softshells, were found in a bald eagle nest in southeast Wisconsin (Watermolen 2004). Nevertheless, due to their highly aquatic lifestyle and swimming abilities, adults are likely not commonly depredated. Wacha and Christiansen (1976, 1977) reported finding four coccidian parasites—single-celled endoparasites—in the bile and intestinal contents of Iowa eastern spiny softshells, though the effect, if any, on the health of turtles was not stated.

Humans pose a significant threat to eastern spiny softshells through overharvesting and climate change. Until recently, commercial harvesting was likely a major threat in Iowa. Based on Iowa Department of Natural Resources records, between 1987 and 2017 commercial turtle trappers harvested an average of 26,281 pounds (11,921 kg) of softshell turtles per year in the state. Harvesters were not required to distinguish between midland smooth and eastern spiny softshells; therefore, it is not possible to know exactly how many eastern spiny softshells were taken. However, the reported average illustrates the magnitude of the softshell catch. Changes to Iowa's turtle harvest regulations in 2017 made a significant difference. In 2021, commercial turtle trappers harvested just 16 pounds (7 kg) of softshell turtles. Recreational harvesters are not required to report their catch; therefore, the number of softshell turtles that they take is unknown, though under current regulations recreational and commercial collection of softshell turtles is closed from May 15 to July 15. Outside of these dates, the rules limit the daily catch to one eastern spiny softshell or one midland smooth softshell for both recreational and commercial harvesters. Passively set fishing nets produced unintentional mortality in softshell turtles in Missouri, leading to the suggestion that certain areas of the Mississippi River be excluded from passive netting and that net designs limiting incidental turtle catches be implemented (Barko and Briggler 2006).

Climate change is an emerging threat that could affect populations in years to come. Anything that increases the frequency and duration of river flooding—such as increased rainfall or increased severity of rain events—may affect nesting success by reducing the availability of suitable nesting habitat and the survival of eggs.

FOOD AND FEEDING. Eastern spiny softshells swim or crawl along the bottom searching under stones or in vegetation for prey. Their cryptic coloration and sedentary bottom-dwelling behavior also make them efficient ambush predators. As prey passes by, the turtle quickly extends its neck and grasps it. It may also simply suck in small animals by rapidly expanding its mouth and pharynx to create a low-pressure area that pulls food into the mouth via a pharyngeal gulp.

These powerful swimmers also chase prey. When captured, small prey is swallowed whole, while the turtle either holds larger prey with its forefeet and tears it into smaller pieces with its jaws or forces it down its throat with its forefeet and swallows it whole.

A. s. spinifera is primarily a carnivorous generalist that consumes a large variety of aquatic insects along with other aquatic invertebrates such as worms, mollusks, and snails as well as fish, larval and adult amphibians, and small snakes (Ernst and Lovich 2009). Crayfish make up a significant proportion of the diet in many locations, including Iowa. Based on stomach content analysis, Williams and Christiansen (1981) concluded that the diet of Iowa eastern spiny softshells was about 55 percent crayfish, which, along with inadvertently swallowed sand and gravel (in 21% of stomachs), suggested a more bottom-foraging lifestyle. Approximately 36.5 percent of the stomachs examined contained large fish probably eaten as carrion, 25 percent contained insects, and only 5.8 percent contained small fish probably eaten alive. Plant material, including acorns and leaves, was found in 61 percent of the stomachs, suggesting that the material had been intentionally eaten.

REPRODUCTION. Males reach sexual maturity at around 3.5 to 4 inches (9–10 cm) straight-line plastron length; females mature later at around 7 to 7.9 inches (18–20 cm) straight-line plastron length (Webb 1962). Ovulation occurs in the spring, likely May or early June, after emergence from overwintering (Robinson and Murphy 1978).

Mating takes place in the water in April or May, shortly after spring emergence. The mating sequence is similar to that of the midland smooth softshell as described by Plummer (1977b). Males actively search for females in the water and at basking sites, investigating the sex of prospective mates by probing under the edges of their carapaces. When a male probes another male, the investigation will either be ignored, or the basking male may confront the probing male with an open-mouthed display of aggression. Females, if unreceptive, may turn and bite the male. If receptive, the female becomes passive as the male mounts her. The male approaches from the rear, crawls onto the female's back, and positions his tail beneath hers, aligning the vents. The male does not grasp the female with his forelimbs but is forced to swim in place atop her to maintain his position. Copulation may last 30 minutes.

Sometime in late June or early July, females lay their eggs, often with many females laying at the same site. Examination of two adult Muscatine County females in the Drake University Research Collection collected the same year found one with oviductal eggs on June 20 and the other with enlarged ovarian follicles but no oviductal eggs on July 18, supporting late June or early July timing of egg laying in southeast Iowa. Females may need to move well outside of their normal activity range to find suitable nesting sites, though the nesting sites themselves are generally located within 328 feet (100 m) of the water. Sandy soil is preferred, but other less suitable soils, such as clay loam, may be used when sand is not available. In southeast Iowa, eastern spiny softshell nests are commonly found on the levees of major rivers, such as the Cedar and Iowa. During his more than 35 years of turtle studies in southeast Iowa, Jim noted that when sandy areas were available beyond the levees, eastern spiny softshells would cross the levees, spend time in oxbows on the landward side of the levees, and then nest in sand dunes beyond the oxbows. At Big Sand Mound Nature Preserve in Muscatine and Louisa Counties, he observed that eastern spiny softshells nested with other turtles in a narrow band about 328 feet (100 m) wide, beginning at the water's edge on the first row of sand dunes adjacent to an oxbow that was separated from the Mississippi River by a large levee.

One midafternoon while checking a drift fence in mid-June at Big Sand Mound Nature Preserve, Jim caught a quick glimpse of a large animal rapidly moving over the crest of a sand dune about 984 feet (300 m) from the water's edge. A short time later, a very large female eastern spiny softshell moving very rapidly cleared the top of a dune about 20 feet (6.1 m) away from him. Without hesitation, she bulldozed over the drift fence and easily beat Jim to the water's edge. Tracks from a second female that stopped at the drift fence as she headed back to the water after nesting led him to where she had nested on a set of dunes approximately 328 feet (100 m) beyond the primary nesting area.

Nesting occurs in the late afternoon or early evening; females dig nests 4 to 10 inches (10–25.4 cm) deep, and they may void their bladder into the chamber to aid in digging. Once the eggs have been laid, the female uses her hind feet to fill in the nest, packing the soil and leaving a small circular depression. Jim observed nesting events in Muscatine County between 4:00 and 7:00 p.m., and Tamplin (2010) observed a female constructing a nest in open sand approximately 39 feet (12 m) from the water at 7:30 p.m. on June 17 in Butler County. Blanchard (1923) found a female in Dickinson County during the middle of the afternoon on July 3 that was "beginning to prepare a nest about twelve feet from the water on a sunny bank, sloping at an angle of about 30 degrees. The impression of the body on the ground showed that the turtle was facing the river, and that the two hind feet were used in digging the double burrow. The latter was already well under way, as it was 4 inches in greatest length and from 1½ to 2 inches deep." The whole nesting sequence is completed quickly. Breckenridge (1960) observed a Minnesota female dig a cavity in 15 minutes, lay a clutch of 17 eggs in 6 minutes, and fill and cover the nest in 5 minutes.

Females lay from 3 to 42 spherical eggs with thick brittle white shells. The record clutch size of 42 eggs was reported by Tamplin (2010) from a nest in Butler County and appears to be the largest known clutch for any North American softshell. Jim and Jeff Parmelee (2003) reported eggs ranging in size from 0.9 to 1.1 inches (2.3–2.9 cm) in diameter for one Iowa nest, and an average egg diameter of 1 inch (2.6 cm) was reported by Tamplin (2010) for 14 nests in Butler County.

Clutch size is proportional to female size, with larger females producing more eggs than smaller females. In the Midwest, clutches of 9 to 25 eggs (Illinois, Cahn 1937), 15 to 38 eggs (Wisconsin, Vogt 1981b), 16 to 42 eggs (Nebraska, Ballinger et al. 2010), and 15 to 20 eggs (Minnesota, Moriarty and Hall 2014) have been reported. In Iowa, Tamplin (2010) reported an average clutch size of 19 eggs for 14 nests (249 total eggs) in Butler County. Counts of enlarged ovarian follicles and shelled oviductal eggs in two Muscatine County females in the Drake University Research Collection resulted in average clutch sizes of 9 eggs (range 5–13; n = 2) and 9 eggs (n = 1), respectively, and LeClere (2013) noted 3 Muscatine County clutches containing 20 to 32 eggs (average = 24). Eastern spiny softshells frequently lay 2 and occasionally 3 clutches per year, although no one has reported 3

Hatchling eastern spiny softshell from Scott County showing the circular markings on the carapace. Photo by Don Becker.

clutches in Iowa to our knowledge. Both of the Muscatine County Drake University Research Collection specimens examined contained a second set of ovarian follicles (n = 13 and 16, respectively), suggesting that they might have produced a second clutch that year.

The eggs hatch from August to early October after approximately 75 to 85 days, depending on incubation temperature, with hatchlings averaging 1.2 to 1.6 inches (3–4 cm) straight-line carapace length. Well-developed embryos close to hatching and recent hatchlings have been shown to make sounds—principally clicks—within the nest (Geller and Casper 2023). The precise function of the sounds is not known, but they may be a form of communication between clutch mates helping to synchronize hatching and nest emergence (Spencer and Janzen 2011). Most hatchlings emerge from the nest in fall, consistent with Christiansen and Gallaway's (1984) report of fall emergence of spiny softshell hatchlings in Louisa County. Unlike many turtle species, eastern spiny softshells have genetic sex determination, with equal numbers of females and males being produced at a wide range of temperatures.

LONGEVITY. Softshell turtles do not establish growth rings as do hard-shelled turtles, so longevity in the wild is uncertain. A wild-caught adult female survived an additional 25 years, 2 months, and 17 days.

GLOSSARY

Anterior: Referring to the forward or head end of an animal.

Barbel: Small whisker-like structures used for smelling, from the Latin *barbula* meaning "little beard."

Buccopharyngeal respiration: Exchange of oxygen and carbon dioxide across the membranes inside the mouth and throat.

Carnivore, carnivorous: Feeding on other animals, including their eggs.

Carrion: The decaying flesh of dead animals.

Cloaca, cloacal: Referring to the common chamber into which the urinary, digestive, and reproductive tracts discharge their contents. Also called the vent.

Corpora lutea: Hormone-secreting structures that develop in an ovary after an ovum has been discharged.

Crepuscular: Active at twilight; dusk and dawn.

Dimorphism, dimorphic: Referring to a condition where sexes of the same species exhibit different morphological characteristics.

Diurnal: Active by day.

Dorsal: Referring to the upper surface or back of an animal.

Ectotherm: An animal that is dependent on external sources of body heat.

Ephemeral: Lasting for a very short time.

Estivation, estivate: Referring to a state of dormancy, similar to hibernation, that takes place in the summer rather than the winter.

Genetic bottleneck: When a population is greatly reduced in size, limiting the genetic diversity of the species because only a small part of the original population survives.

Genus (plural genera): Principal taxonomic category that ranks above a species and below a family.

Great Plains: A large expanse of land, much of it grassland, located east of the Rocky Mountains and west of the Missouri River that encompasses all or parts of Montana, North Dakota, South Dakota, Wyoming, Nebraska, Kansas, Colorado, Oklahoma, Texas, and New Mexico.

Growth rings: A series of grooves created on the carapace or plastron of a turtle during overwintering that can be used to estimate age in some species.

Herpetology, herpetologist, herpetological, herpetofauna, herpetofaunal: Referring to the study of amphibians and reptiles.

Hibernacula: Winter quarters of an overwintering animal.

Hybridization: Mating between two different species.

Inbreeding depression: The reduced survival and fertility of offspring of related individuals because of the accumulation of deleterious genes.

Melanism, melanistic: Referring to an increased amount of black or nearly black pigmentation in the skin or scales of an individual.

Mesopredator: Midranking predator in a food chain, such as a raccoon, fox, skunk, or snake, that typically preys on smaller animals. Mesopredators often increase in abundance when larger predators are eliminated.

Morphological: Referring to the study of the shapes and structure of plants and animals.

Omnivore, omnivorous: Feeding on food of both plant and animal origin.

Phylogenetic: Relating to the evolutionary development and diversification of a species or group of organisms or of a particular feature of an organism.

Posterior: Referring to the rear end of an animal.

Sexual dimorphism: Distinct difference in size or appearance between the sexes.

Species: The basic unit of classification and a taxonomic rank of an organism; a group of organisms that can reproduce with one another in nature and produce fertile offspring.

Subspecies: Taxonomic group that is a division of a species; designates a population of a particular geographic region genetically distinguishable from other populations of the same species and capable of interbreeding with them where its range overlaps theirs.

Temperature-dependent sex determination: Temperature of the egg during embryonic development that determines the sex of the offspring.

Terrapin: A group of several species of small turtles that live in fresh or brackish water.

Thermoregulation, thermoregulator: The maintenance or regulation of body temperature. Because turtles are cold-blooded—ectothermic—they use external sources of temperature to regulate their body temperatures.

Tortoise: A turtle that lives on land.

Vent: The external orifice of the cloaca.

Ventral: Referring to the underside or belly of an animal.

Vitellogenesis, vitellogenic: Referring to the process of yolk formation via nutrients being deposited in the immature egg.

Winterkill: Mortality of aquatic organisms in a waterbody during winter due to low levels of dissolved oxygen resulting when the waterbody freezes over and is covered by layers of snow. May also happen to terrestrial turtles if they are caught above ground after the soil freezes, preventing them from burrowing in for the winter.

Xeric: Containing little moisture; very dry.

REFERENCES

Ackerman, R. A., T. A. Rimkus, and D. B. Lott. 2008. Water relations of snapping turtle eggs. Pp. 133–145 in A. C. Steyermark, M. S. Finkler, and R. J. Brooks (eds.), Biology of the Snapping Turtle (*Chelydra serpentina*). Johns Hopkins University Press, Baltimore.

Agassiz, L. 1857. Contributions to the Natural History of the United States of America. Vol. 1. Little, Brown and Company, Boston.

Alderman, S. L., C. L. Riggs, O. M. Bullingham, T. E. Gillis, and D. E. Warren. 2021. Cold acclimation induces life stage–specific responses in the cardiac proteome of western painted turtles (*Chrysemys picta bellii*): Implications for anoxia tolerance. Journal of Experimental Biology 224:jeb242387.

Alibardi, L., and M. Toni. 2006. Skin structure and cornification proteins in the soft-shelled turtle *Trionyx spiniferus*. Zoology 109:182–195.

Anderson, P. 1965. The Reptiles of Missouri. University of Missouri Press, Columbia.

Anderson, P. K. 1958. The photic responses and water-approach behavior of hatchling turtles. Copeia 1958:211–215.

Anderson, R. V., M. L. Gutierrez, and M. A. Romano. 2002. Turtle habitat use in a reach of the upper Mississippi River. Journal of Freshwater Ecology 17:171–177.

Anonymous. 1881. History of Fremont County, Iowa. Iowa Historical Company, Des Moines.

Aresco, M. J., M. A. Ewert, M. S. Gunzburger, G. L. Heinrich, and P. A. Meylan. 2006. *Chelydra serpentina* – Snapping turtle. Chelonian Research Monographs 3:44–57.

Bailey, R. M. 1941. The occurrence of the wood turtle in Iowa. Copeia 1941:265.

Baker, P. J., J. P. Costanzo, J. B. Iverson, and R. E. Lee. 2003. Adaptations to terrestrial overwintering of hatchling northern map turtles, *Graptemys geographica*. Journal of Comparative Physiology B 173:643–651.

Baker, R. E., and J. C. Gillingham. 1983. An analysis of courtship behavior in Blanding's turtle, *Emydoidea blandingi*. Herpetologica 39:166–173.

Ballinger, R. E., J. D. Lynch, and G. R. Smith. 2010. Amphibians and Reptiles of Nebraska. University of Nebraska Press, Lincoln.

Barko, V. A., and J. T. Briggler. 2006. Midland smooth softshell (*Apalone mutica*) and spiny softshell (*Apalone spinifera*) turtles in the middle Mississippi River: Habitat associations, population structure, and implications for conservation. Chelonian Conservation and Biology 5:225–231.

Behler, J. L., and C. M. Castellano. 2005. *Glyptemys insculpta* (North American wood turtle): Nest invading plants. Herpetological Review 36:311.

Belkin, D. A., and C. Gans. 1968. An unusual chelonian feeding niche. Ecology 49:768–769.

Beltz, E. 2007. Scientific and common names of the reptiles and amphibians of North America – Explained. Available at http://ebeltz.net/herps/etymain.html.

Ben-Ezra, E., G. Bulté, and G. Blouin-Demers. 2008. Are locomotor performances coadapted to preferred basking temperature in the northern map turtle (*Graptemys geographica*)? Journal of Herpetology 42:322–331.

Bennett, A. M., M. Keevil, and J. D. Litzgus. 2010. Spatial ecology and population genetics of northern map turtles (*Graptemys geographica*) in fragmented and continuous habitats in Canada. Chelonian Conservation and Biology 9:185–195.

Bennett, A. M., and J. D. Litzgus. 2014. Injury rates of freshwater turtles on a recreational waterway in Ontario, Canada. Journal of Herpetology 48:262–266.

Bentley, P. J., and K. Schmidt-Nielsen. 1966. Cutaneous water loss in reptiles. Science 151:1547–1549.

Bernstein, N. P., and R. W. Black. 2005. Thermal environment of overwintering ornate box turtles, *Terrapene ornata ornata*, in Iowa. American Midland Naturalist 153:370–377.

Bernstein, N. P., and J. L. Christiansen. 2011. Response of a yellow mud turtle (*Kinosternon flavescens* Agassiz) community to habitat change: Management implication for a nature preserve. Natural Areas Journal 31:414–419.

Bernstein, N. P., R. H. Fendrich, and S. A. McCollum. 2023. Do home range, movement patterns, and habitat use of ornate box turtles (*Terrapene ornata ornata*) differ among age classes? Journal of Herpetology 57:1–10.

Bernstein, N. P., S. A. McCollum, and R. W. Black. 2015. How do predators locate nests of ornate box turtles (*Terrapene ornata*)? A field experiment. Herpetological Conservation and Biology 10:44–53.

Bernstein, N. P., S. A. McCollum, T. J. VanDeWalle, and D. F. Hughes. 2023. Longevity estimates of ornate box turtles (*Terrapene ornata*) in Iowa. Chelonian Conservation and Biology 22:220–224.

Bernstein, N. P., R. J. Richtsmeier, R. W. Black, and B. R. Montgomery. 2007. Home range and philopatry in the ornate box turtle, *Terrapene ornata ornata*, in Iowa. American Midland Naturalist 157:162–174.

Bernstein, N. P., R. G. Todd, M. Y. Baloch, S. A. McCollum, T. Skorczewski, K. A. Mickael, and J. E. Eastham. 2018. Morphometric models of growth in ornate box turtles (*Terrapene ornata ornata*) as related to growth rings. Chelonian Conservation and Biology 17:197–205.

Berry, J. F., and C. M. Berry. 1984. A reanalysis of geographic variation and systematics in the yellow mud turtle, *Kinosternon flavescens* (Agassiz). Annals of the Carnegie Museum 53:185–206.

Bickham, J. W., M. S. Springer, and B. J. Gallaway. 1984. Distributional survey of the yellow mud turtle (*Kinosternon flavescens*) in Iowa, Illinois, and Missouri: A proposed endangered species. Southwestern Naturalist 29:123–132.

Blair, W. F. 1976. Some aspects of the biology of the ornate box turtle, *Terrapene ornata*. Southwestern Naturalist 21:89–103.

Blanchard, F. N. 1923. The amphibians and reptiles of Dickinson County, Iowa. University of Iowa Studies in Natural History 10:19–26.

Blasco, R., J. Rosell, K. T. Smith, L. C. Maul, P. Sañudo, R. Barkai, and A. Gopher. 2016. Tortoises as a dietary supplement: A view from the Middle Pleistocene site of Qesem Cave, Israel. Quaternary Science Reviews 133:165–182.

Bluett, R. D., W. E. Louis, D. A. Newhouse, C. J. Handel, Jr., and J. H. Kube. 2013. Longitudinal structuring of turtle assemblages in an altered river in central Illinois, USA: Implications for conservation. Transactions of the Illinois State Academy of Science 106:47–53.

Bodie, J. R., and R. D. Semlitsch. 2000. Spatial and temporal use of floodplain habitats by lentic and lotic species of aquatic turtles. Oecologia 122:138–146.

Bowen, K. D., and F. J. Janzen. 2005. Rainfall and depredation of nests of the painted turtle, *Chrysemys picta*. Journal of Herpetology 39:649–652.

Braun, A. P., and Q. E. Phelps. 2016. Habitat use by five turtle species in the middle Mississippi River. Chelonian Conservation and Biology 15:62–68.

Braun, D. R., J. W. Harris, N. E. Levin, J. T. McCoy, A. I. Herries, M. K. Bamford, L. C. Bishop, B. G. Richmond, and M. Kibunjia. 2010. Early hominin diet included diverse terrestrial and aquatic animals 1.95 Ma in East Turkana, Kenya. Proceedings of the National Academy of Sciences 107:10002–10007.

Brecke, B., and J. J. Moriarty. 1989. *Emydoidea blandingii* (Blanding's turtle): Longevity. Herpetological Review 20:53.

Breckenridge, W. J. 1960. A spiny soft-shell turtle nest study. Herpetologica 16:284–285.

Briggler, J. T., and T. R. Johnson. 2021. The Amphibians and Reptiles of Missouri. 3rd ed. Missouri Department of Conservation, Jefferson City.

Britson, C. A., and W. H. Gutzke. 1993. Antipredator mechanisms of hatchling freshwater turtles. Copeia 1993:435–440.

Brown, D. J., and C. N. Lapin. 2022. Wood turtle, *Glyptemys insculpta* (LeConte 1830). Pp. 768–783 in J. M. Kapfer and D. J. Brown (eds.), Amphibians and Reptiles of Wisconsin. University of Wisconsin Press, Madison.

Brown, D. J., M. D. Nelson, D. J. Rugg, R. R. Buech, and D. M. Donner. 2016. Spatial and temporal habitat-use patterns of wood turtles at the western edge of their distribution. Journal of Herpetology 50:347–356.

Brown, G. P., and R. J. Brooks. 1994. Characteristics of and fidelity to hibernacula in a northern population of snapping turtles, *Chelydra serpentina*. Copeia 1994:222–226.

Brown, L. E., and D. Moll. 1979. The status of the nearly extinct Illinois mud turtle (*Kinosternon flavescens spooneri* Smith 1951) with recommendations for its conservation. Milwaukee Public Museum Special Publications in Biology and Geology 3.

Brumfiel, D. M. 1919. The animal ecology of Johnson County. University of Iowa Laboratories of Natural History Bulletin 8:1–37.

Buikstra, J. E., and G. R. Milner. 1991. Isotopic and archaeological interpretations of diet in the central Mississippi valley. Journal of Archaeological Science 18:319–329.

Bull, J. J., and R. C. Vogt. 1979. Temperature-dependent sex determination in turtles. Science 206:1186–1188.

Bull, J. J., and R. C. Vogt. 1981. Temperature-sensitive periods of sex determination in emydid turtles. Journal of Experimental Zoology 218:435–440.

Bulté, G., R. R. Germain, C. M. O'Connor, and G. Blouin-Demers. 2013. Sexual dichromatism in the northern map turtle, *Graptemys geographica*. Chelonian Conservation and Biology 12:187–192.

Bulté, G., D. J. Irschick, and G. Blouin-Demers. 2008. The reproductive role hypothesis explains trophic morphology dimorphism in the northern map turtle. Functional Ecology 22:824–830.

Burken, R. R. 1978. Aspects of the Niche of the Common Snapping Turtle, *Chelydra serpentina serpentina*, in Iowa. Master's thesis, Drake University, Des Moines.

Butler, C. J., B. D. Stanila, J. B. Iverson, P. A. Stone, and M. Bryson. 2016. Projected changes in climatic suitability for *Kinosternon* turtles by 2050 and 2070. Ecology and Evolution 6:7690–7705.

Cadi, A., and P. Joly. 2004. Impact of the introduction of the red-eared slider (*Trachemys scripta elegans*) on survival rates of the European pond turtle (*Emys orbicularis*). Biodiversity and Conservation 13:2511–2518.

Cagle, F. R. 1942. Herpetological fauna of Jackson and Union Counties, Illinois. American Midland Naturalist 28:164–200.

Cagle, F. R. 1944a. Home range, homing behavior, and migration in turtles. Miscellaneous Publications of the Museum of Zoology of the University of Michigan 61:1–34.

Cagle, F. R. 1944b. Sexual maturity in the female of the turtle *Pseudemys scripta elegans*. Copeia 1944:149–152.

Cagle, F. R. 1948. Sexual maturity in the male turtle, *Pseudemys scripta troostii*. Copeia 1948:108–111.

Cagle, F. R. 1950. The life history of the slider turtle, *Pseudemys scripta troostii* (Holbrook). Ecological Monographs 20:31–54.

Cagle, F. R. 1953. Two new subspecies of *Graptemys pseudogeographica*. Occasional Papers of the Museum of Zoology of the University of Michigan 546:1–17.

Cahn, A. R. 1937. The Turtles of Illinois. University of Illinois, Urbana.

Carlander, K. D., and R. B. Moorman. 1951a. Iowa's land turtles. Iowa Conservationist 10:113, 119.

Carlander, K. D., and R. B. Moorman. 1951b. Pond and river turtles. Iowa Conservationist 10:125.

Carr, A. 1952. Handbook of Turtles: The Turtles of the United States, Canada, and Baja California. Cornell University Press, Ithaca.

Carr, A. F., Jr. 1938. Notes on the *Pseudemys scripta* complex. Herpetologica 1:131–135.

Carrière, M. A., and G. Blouin-Demers. 2010. Habitat selection at multiple spatial scales in northern map turtles (*Graptemys geographica*). Canadian Journal of Zoology 88:846–854.

Carrière, M. A., N. Rollinson, A. N. Suley, and R. J. Brooks. 2008. Thermoregulation when the growing season is short: Sex-biased basking patterns in a northern population of painted turtles (*Chrysemys picta*). Journal of Herpetology 42:206–209.

Casper, G. S. 1998. Movement patterns, habitat use, and conservation strategies in a small population of Blanding's turtle (*Emydoidea blandingii*) in southern Wisconsin. Abstract, Turtle Workshop, May 7–8, J. F. Bell Museum of Natural History, Minneapolis.

Catania, K. C. 2008. Worm grunting, fiddling, and charming: Humans unknowingly mimic a predator to harvest bait. PLOS ONE 3:e3472.

Center for Biological Diversity. 2009. Eight states petitioned to end unsustainable turtle harvest: Turtles contaminated with mercury and other toxins sold as food. Available at https://www.biologicaldiversity.org/news/press_releases/2009/freshwater-turtles-03-11-2009.html.

Christiansen, J. L. 1961. The amphibians and reptiles of Buena Vista County, Iowa. Proceedings of the Iowa Academy of Science 68:625–634.

Christiansen, J. L. 1981. Population trends among Iowa's amphibians and reptiles. Proceedings of the Iowa Academy of Science 88:24–27.

Christiansen, J. L. 1998. Perspectives on Iowa's declining amphibians and reptiles. Journal of the Iowa Academy of Science 105:109–114.

Christiansen, J. L. 2001. Non-native amphibians and reptiles in Iowa. Journal of the Iowa Academy of Science 108:210–211.

Christiansen, J. L., and R. M. Bailey. 1988. The lizards and turtles of Iowa. Iowa Department of Natural Resources Nongame Technical Series 3:1–20.

Christiansen, J. L., and R. M. Bailey. 1997. The lizards and turtles of Iowa (revised). Iowa Department of Natural Resources Nongame Technical Series 3:1–20.

Christiansen, J. L., N. P. Bernstein, and E. R. Jacobson. 2021. A shell disease in an Iowa, USA population of yellow mud turtles, *Kinosternon flavescens*: Evidence for an isolated local distribution. Herpetological Review 52:293–298.

Christiansen, J. L., N. P. Bernstein, C. A. Phillips, J. T. Briggler, and D. Kangas. 2012. Declining populations of yellow mud turtles (*Kinosternon flavescens*) in Iowa, Illinois, and Missouri. Southwestern Naturalist 57:304–313.

Christiansen, J. L., and J. W. Bickham. 1989. Possible historic effects of pond drying and winterkill on the behavior of *Kinosternon flavescens* and *Chrysemys picta*. Journal of Herpetology 23:91–94.

Christiansen, J. L., and R. R. Burken. 1978. The endangered and uncommon amphibians and reptiles of Iowa. Iowa Science Teachers Journal 15:1–86.

Christiansen, J. L., and R. R. Burken. 1979. Growth and maturity of the snapping turtle (*Chelydra serpentina*) in Iowa. Herpetologica 35:261–266.

Christiansen, J. L., J. A. Cooper, and J. W. Bickham. 1984. Reproduction of *Kinosternon flavescens* in Iowa. Southwestern Naturalist 29:349–351.

Christiansen, J. L., J. A. Cooper, J. W. Bickham, B. J. Galloway, and M. A. Springer. 1985. Aspects of the natural history of the yellow mud turtle, *Kinosternon flavescens*

(Kinosternidae) in Iowa: A proposed endangered species. Southwestern Naturalist 30:413–425.

Christiansen, J. L., and A. E. Dunham. 1972. Reproduction of the yellow mud turtle (*Kinosternon flavescens flavescens*) in New Mexico. Herpetologica 28:130–137.

Christiansen, J. L., and B. J. Gallaway. 1984. Raccoon removal, nesting success, and hatchling emergence in Iowa turtles with special reference to *Kinosternon flavescens* (Kinosternidae). Southwestern Naturalist 29:343–348.

Christiansen, J. L., B. J. Gallaway, and J. W. Bickham. 1990. Population estimates and geographic distribution of the yellow mud turtle (*Kinosternon flavescens*) in Iowa. Journal of the Iowa Academy of Science 97:105–108.

Christiansen, J. L., J. M. Grzybowski, and R. A. Kodama. 1996. Melanomacrophage aggregations and their age relationships in the yellow mud turtle, *Kinosternon flavescens* (Kinosternidae). Pigment Cell Research 9:185–190.

Christiansen, J. L., J. M. Grzybowski, and B. P. Rinner. 2004. Facial lesions in turtles, observations of prevalence, reoccurrence, and multiple origins. Journal of Herpetology 37:293–298.

Christiansen, J. L., and C. M. Mabry. 1985. The amphibians and reptiles of Iowa's Loess Hills. Proceedings of the Iowa Academy of Science 92:159–163.

Christiansen, J. L., and E. O. Moll. 1973. Latitudinal reproductive variation within a single subspecies of painted turtle, *Chrysemys picta bellii*. Herpetologica 29:152–163.

Christiansen, J. L., and J. R. Parmelee. 2003. A key to the eggs and nests of Iowa turtles. Journal of the Iowa Academy of Science 110:44–50.

Christiansen, J. L., and T. J. VanDeWalle. 1999. Distribution and age structure of Blanding's turtles in northern Iowa. Unpublished report submitted to the Iowa Department of Natural Resources, Des Moines.

Christiansen, J. L., and T. VanDeWalle. 2000. Effectiveness of three trap types in drift fence surveys. Herpetological Review 31:158–160.

Clark, D. B., and J. W. Gibbons. 1969. Dietary shift in the turtle *Pseudemys scripta* (Schoepff) from youth to maturity. Copeia 1969:704–706.

Clark, H. W., and J. B. Southall. 1920. Fresh-water turtles: A source of meat supply. U.S. Government Printing Office, Washington, D.C.

Close, L. M., and R. A. Seigel. 1997. Differences in body size among populations of red-eared sliders (*Trachemys scripta elegans*) subjected to different levels of harvesting. Chelonian Conservation and Biology 2:563–566.

Cochran, P. A. 1987. *Graptemys geographica* (map turtle): Adult mortality. Herpetological Review 18:37.

Collison, P. J., and J. F. Doershuk. 2021. Vertebrate faunal remains from the Mini-Wakan site: A preliminary report. Journal of the Iowa Archaeological Society 68:48–58.

Conant, R. 1938. The reptiles of Ohio. American Midland Naturalist 20:1–200.

Conant, R. 1975. A Field Guide to Reptiles and Amphibians of Eastern and Central North America. Peterson Field Guide Series. Houghton Mifflin Company, Boston.

Congdon, J. D., G. L. Breitenbach, R. C. van Loben Sels, and D. W. Tinkle. 1987. Reproduction and nesting ecology of snapping turtles (*Chelydra serpentina*) in southeastern Michigan. Herpetologica 43:39–54.

Congdon, J. D., S. W. Gotte, and R. W. McDiarmid. 1992. Ontogenetic changes in habitat use by juvenile turtles (*Chelydra serpentina* and *Chrysemys picta*). Canadian Field Naturalist 106:241–248.

Congdon, J. D., R. D. Nagle, O. M. Kinney, M. Osenioski, H. W. Avery, R. C. van Loben Sels, and D. W. Tinkle. 2000. Nesting ecology and embryo mortality: Implications for hatchling success and demography of Blanding's turtles (*Emydoidea blandingii*). Chelonian Conservation and Biology 3:569–579.

Congdon, J. D., D. W. Tinkle, G. L. Breitenbach, and R. C. van Loben Sels. 1983. Nesting ecology and hatching success in the turtle *Emydoidea blandingi*. Herpetologica 39:417–429.

Congdon, J. D., and R. C. van Loben Sels. 1993. Relationships of reproductive traits and body size with attainment of sexual maturity and age in Blanding's turtles (*Emydoidea blandingi*). Journal of Evolutionary Biology 6:547–557.

Conner, C. A., B. A. Douthitt, and T. J. Ryan. 2005. Descriptive ecology of a turtle assemblage in an urban landscape. American Midland Naturalist 153:428–435.

Converse, S. J., J. B. Iverson, and J. A. Savidge. 2002. Activity, reproduction and overwintering behavior of ornate box turtles (*Terrapene ornata ornata*) in the Nebraska Sandhills. American Midland Naturalist 148:416–422.

Cooper, J. 1975. Behavioral Aspects of the Life History of the Illinois Mud Turtle, *Kinosternon flavescens spooneri*. Master's thesis, Drake University, Des Moines.

Cooper, J. 1977. Vest-pocket turtle. Natural History 86:52–57.

Cosentino, B. J., R. L. Schooley, and C. A. Phillips. 2010. Wetland hydrology, area, and isolation influence occupancy and spatial turnover of the painted turtle, *Chrysemys picta*. Landscape Ecology 25:1589–1600.

Costanzo, J. P., P. J. Baker, and R. E. Lee, Jr. 2006. Physiological responses to freezing in hatchlings of freeze-tolerant and -intolerant turtles. Journal of Comparative Physiology B 176:697–707.

Costanzo, J. P., J. B. Iverson, M. F. Wright, and R. E. Lee, Jr. 1995. Cold hardiness and overwintering strategies in an assemblage of northern turtles. Ecology 76:1772–1785.

Costanzo, J. P., J. D. Litzgus, J. L. Larson, J. B. Iverson, and R. E. Lee, Jr. 2001. Cold-hardiness and evaporative water loss in hatchling turtles. Physiological Biochemical Zoology 74:510–519.

Coulter, M. W. 1957. Predation by snapping turtles upon aquatic birds in Maine marshes. Journal of Wildlife Management 21:17–21.

Cowin, S., and J. Cebek. 2006. Feasibility of using plastron markings in young wood turtles (*Glyptemys insculpta*) as a technique for identifying individuals. Herpetological Review 37:305–306.

Crawford, K. M. 1991. The winter environment of painted turtles, *Chrysemys picta*:

Temperature, dissolved oxygen, and potential cues for emergence. Canadian Journal of Zoology 69:2493–2498.

Crawford, K. M. 1994. Patterns of energy substrate utilization in overwintering painted turtles, *Chrysemys picta*. Comparative Biochemistry and Physiology Part A: Physiology 109:495–502.

Crocker, C. E., T. E. Graham, G. R. Ultsch, and D. C. Jackson. 2000. Physiology of common map turtles (*Graptemys geographica*) hibernating in the Lamoille River, Vermont. Journal of Experimental Zoology 288:143–148.

Davis, J. B., B. D. Leopold, R. M. Kaminski, and R. R. Cox. 2009. Wood duck duckling mortality and habitat implications in floodplain systems. Wetlands 29:607–614.

Dodd, C. K., Jr. 1983. A review of the status of the Illinois mud turtle *Kinosternon flavescens spooneri* Smith. Smithsonian Biological Conservation 27:141–156.

Dodd, C. K., Jr. 2001. North American Box Turtles: A Natural History. University of Oklahoma Press, Norman.

Dodge, C., and L. S. Miller. 1955. The yellow mud turtle, *Kinosternon flavescens spooneri* Smith, in Iowa. Natural History Miscellanea 144:1–3.

Dodge, C. H. 1956a. The musk turtle in Iowa. Herpetologica 12:176.

Dodge, C. H. 1956b. An unusually large Blanding's turtle from Iowa. Herpetologica 12:328.

Dolan, B. J. 2007. Exploring Okoboji Oneota: Analysis and Regional Comparison of Faunal Data from Gillett Grove (13CY2), Clay County, Iowa. Master's thesis, University of Nebraska, Lincoln.

DonnerWright, D. M., M. A. Bozek, J. R. Probst, and E. M. Anderson. 1999. Responses of turtle assemblage to environmental gradients in the St. Croix River in Minnesota and Wisconsin, USA. Canadian Journal of Zoology 77:989–1000.

Dorland, A., T. Rytwinski, and L. Fahrig. 2014. Do roads reduce painted turtle (*Chrysemys picta*) populations? PLOS ONE 9:e98414.

Doroff, A. M., and L. B. Keith. 1990. Demography and ecology of an ornate box turtle (*Terrapene ornata*) population in south-central Wisconsin. Copeia 1990:387–399.

Dreslik, M. J. 1999. Dietary notes on the red-eared slider (*Trachemys scripta*) and river cooter (*Pseudemys concinna*) from southern Illinois. Transactions of the Illinois State Academy of Science 92:233–241.

Dreslik, M. J., and A. R. Kuhns. 2000. Early season basking in the red-eared slider, *Trachemys scripta*. Transactions of the Illinois State Academy of Science 93:215–220.

Dreslik, M. J., A. R. Kuhns, and C. A. Phillips. 2005. Structure and composition of a southern Illinois freshwater turtle assemblage. Northeastern Naturalist 12:173–186.

Dreslik, M. J., and C. A. Phillips. 2005. Turtle communities in the Upper Midwest, USA. Journal of Freshwater Ecology 20:149–164.

Dubois, Y., G. Blouin-Demers, B. Shipley, and D. Thomas. 2009. Thermoregulation and habitat selection in wood turtles *Glyptemys insculpta*: Chasing the sun slowly. Journal of Animal Ecology 78:1023–1032.

Dunson, W. A. 1960. Aquatic respiration in *Trionyx spinifer asper*. Herpetologica 16:277–283.

Edmonds, D. A., L. Adamovicz, M. C. Allender, A. Colton, R. Nÿboer, and M. Dreslik. 2023. Evaluating population persistence of ornate box turtles (*Terrapene ornata*) at the northeast edge of their distribution. Wildlife Biology e01183.

Edmonds, D. A., L. Adamovicz, M. C. Allender, and M. J. Dreslik. 2019. Reproductive output of ornate box turtles (*Terrapene ornata*) in Illinois, USA. Herpetological Conservation and Biology 15:467–475.

Edmonds, D. A., E. M. Bach, A. L. Colton, I. S. Jaquet, E. J. Kessler, and M. J. Dreslik. 2023. Avoiding mortality: Timing of prescribed burns in ornate box turtle habitat. Journal of Wildlife Management 88, doi:10.1002/jwmg.22510.

Edwards, A. L., and G. Blouin-Demers. 2007. Thermoregulation as a function of thermal quality in a northern population of painted turtles, *Chrysemys picta*. Canadian Journal of Zoology 85:526–535.

Eisner, T., W. E. Conner, K. Hicks, K. R. Dodge, H. I. Rosenberg, T. H. Jones, M. Cohen, and J. Meinwald. 1977. Stink of stinkpot turtle identified: ω-phenylalkanoic acids. Science 196:1347–1349.

Ernst, C. H. 1972. Temperature-activity relationship in the painted turtle, *Chrysemys picta*. Copeia 1972:217–222.

Ernst, C. H. 1986. Ecology of the turtle, *Sternotherus odoratus,* in southeastern Pennsylvania. Journal of Herpetology 20:341–352.

Ernst, C. H., and E. M. Ernst. 1973. Biology of *Chrysemys picta bellii* in southwestern Minnesota. Journal of the Minnesota Academy of Science 38:77–80.

Ernst, C. H., and J. E. Lovich. 2009. Turtles of the United States and Canada. Johns Hopkins University Press, Baltimore.

Errington, P. L. 1941. Versatility in feeding and population maintenance of the muskrat. Journal of Wildlife Management 5:68–89.

Feltz, J., and J. Tamplin. 2007. Effect of substrate on selected temperature in juvenile spiny softshell turtles (*Apalone spinifera*). Chelonian Conservation and Biology 6:177–184.

Filoramo, N. I., and F. J. Janzen. 2002. An experimental study of the influence of embryonic water availability, body size, and clutch on survivorship of neonatal red-eared sliders, *Trachemys scripta elegans*. Herpetologica 58:67–74.

Fitch, H. S. 1958. Home ranges, territories, and seasonal movements of vertebrates of the University of Kansas Natural History Reservation. University of Kansas Museum of Natural History 11:63–326.

Fitch, H. S., and M. V. Plummer. 1975. A preliminary ecological study of the soft-shelled turtle *Trionyx muticus* in the Kansas River. Israel Journal of Zoology 24:28–42.

Flanagan, M., V. Roy-McDougall, G. Forbes, and G. Forbes. 2013. Survey methodology for the detection of wood turtles (*Glyptemys insculpta*). Canadian Field-Naturalist 127:216–223.

Ford, D. K., and D. Moll. 2004. Sexual and seasonal variation in foraging patterns in the stinkpot, *Sternotherus odoratus,* in southwestern Missouri. Journal of Herpetology 38:296–301.

Frankson, R., K. E. Kunkel, S. M. Champion, and J. Runkle. 2022. Iowa state climate summary 2022. NOAA Technical Report NESDIS 150-IA. NOAA/NESDIS, Silver Spring, Md.

Freedberg, S., and E. M. Myers. 2012. Cytonuclear equilibrium following interspecific introgression in a turtle lacking sex chromosomes. Biological Journal of the Linnean Society 106:405–417.

Frese, P. W. 2003. Tallgrass prairie amphibian and reptile assemblage: Fire mortality. Herpetological Review 34:159–160.

Fritz, U., and P. Havaš. 2013. Order Testudines: 2013 update. Zootaxa 3703:12–14.

Fritz, U., and P. Havaš. 2014. On the reclassification of box turtles (Terrapene): A response to Martin et al. (2014). Zootaxa 3835:295–298.

Fuselier, L., and D. Edds. 1994. Habitat partitioning among three sympatric species of map turtles, genus *Graptemys*. Journal of Herpetology 28:154–158.

Galois, P., M. Léveillé, L. Bouthillier, C. Daigle, and S. Parren. 2002. Movement patterns, activity and home range of the eastern spiny softshell turtle (*Apalone spinifera*) in northern Lake Champlain, Québec, Vermont. Journal of Herpetology 36:402–411.

Gamble, T., and A. M. Simmons. 2003. The commercial harvest of painted turtles in Minnesota. Final report to the Minnesota Department of Natural Resources, Natural Heritage and Nongame Research Program.

Gamble, T., and A. M. Simmons. 2004. Comparison of harvested and nonharvested painted turtle populations. Wildlife Society Bulletin 32:1269–1277.

Geller, G. A. 2012. Notes on the nest predation dynamics of *Graptemys* at two Wisconsin sites using trail camera monitoring. Chelonian Conservation and Biology 11:197–205.

Geller, G. A. 2015. A test of substrate sweeping as a strategy to reduce raccoon predation of freshwater turtle nests, with insights from supplemental artificial nests. Chelonian Conservation and Biology 14:64–72.

Geller, G. A. 2020. *Graptemys ouachitensis* (Ouachita map turtle): Predation. Herpetological Review 51:314–315.

Geller, G. A., and G. S. Casper. 2019a. *Chelydra serpentina* (snapping turtle): Hatchling sounds. Herpetological Review 50:768–769.

Geller, G. A., and G. S. Casper. 2019b. Late term embryos and hatchlings of Ouachita map turtles (*Graptemys ouachitensis*) make sounds within the nest. Herpetological Review 50:449–452.

Geller, G. A., and G. S. Casper. 2023. Sound production by late-term embryos and hatchlings of the spiny softshell turtle, *Apalone spinifera* (LeSueur, 1827), before nest exit. Herpetology Notes 16:483–491.

Geller, G. A., G. S. Casper, and B. J. Halstead. 2020a. Dispersal of hatchling Ouachita map turtles (*Graptemys ouachitensis*) from natural nests on the lower Wisconsin River, Wisconsin. Chelonian Conservation and Biology 19:236–245.

Geller, G. A., G. S. Casper, and B. J. Halstead. 2020b. Hatchling emergence ecology of

Ouachita map turtles (*Graptemys ouachitensis*) on the lower Wisconsin River, Wisconsin. Chelonian Conservation and Biology 19:217–235.

Gibbons, J. W. 1987. Why do turtles live so long? BioScience 37:262–269.

Gibbons, J. W. 2013. A long-term perspective of delayed emergence (aka overwintering) in hatchling turtles: Some they do and some they don't, and some you just can't tell. Journal of Herpetology 47:203–214.

Gibbons, J. W., J. L. Greene, and J. D. Congdon. 1983. Drought-related responses of aquatic turtle populations. Journal of Herpetology 17:242–246.

Gibbons, J. W., D. E. Scott, T. J. Ryan, K. A. Buhlmann, T. D. Tuberville, B. S. Metts, J. L. Greene, T. Mills, Y. Leiden, S. Poppy, and C. T. Winne. 2000. The global decline of reptiles, déjà vu amphibians: Reptile species are declining on a global scale. Six significant threats to reptile populations are habitat loss and degradation, introduced invasive species, environmental pollution, disease, unsustainable use, and global climate change. BioScience 50:653–666.

Gibbons, J. W., and R. D. Semlitsch. 1982. Survivorship and longevity of a long-lived vertebrate species: How long do turtles live? Journal of Animal Ecology 51:523–527.

Goldsmith, W. M. 1944 [1945]. Notes on the egg laying habits of the soft shell turtles. Proceedings of the Iowa Academy of Science 51:447–449.

Graham, T. E., and A. A. Graham. 1991. *Trionyx spiniferus spiniferus* (eastern spiny softshell): Burying behavior. Herpetological Review 22:56–57.

Graham, T. E., and A. A. Graham. 1992. Metabolism and behavior of wintering common map turtles, *Graptemys geographica,* in Vermont. Canadian Field-Naturalist 106:517–519.

Greaves, W. F., and J. D. Litzgus. 2007. Overwintering ecology of wood turtles (*Glyptemys insculpta*) at the species' northern range limit. Journal of Herpetology 41:32–40.

Gritters, S., K. Osterkamp, D. Weiss, C. Dolan, M. Mason, K. Kinkead, and M. Steuck. 2013. Iowa's Commercial Turtle Harvest. Iowa Department of Natural Resources, Des Moines.

Gritters, S. A., and L. M. Mauldin. 1994. Four years of turtle collections on Navigation Pool 13 of the upper Mississippi River. Report to the National Biological Survey, Environmental Management Technical Center, Onalaska, Wisconsin. LTRMP 94-S010.

Grobman, A. B. 1990. The effect of soil temperatures on emergence from hibernation of *Terrapene carolina* and *T. ornata*. American Midland Naturalist 124:366–371.

Hammer, D. A. 1969. Parameters of a marsh snapping turtle population Lacreek Refuge, South Dakota. Journal of Wildlife Management 33:995–1005.

Harding, J. H., and T. J. Bloomer. 1979. The wood turtle, *Clemmys insculpta* . . . a natural history. Bulletin of the New York Herpetological Society 15:9–26.

Harding, J. H., and D. A. Mifsud. 2017. Amphibians and Reptiles of the Great Lakes Region. Rev. ed. University of Michigan Press, Ann Arbor.

Harlan, J. R., E. B. Speaker, and J. Mayhew. 1987. Iowa Fish and Fishing. Iowa Department of Natural Resources, Des Moines.

Harless, M. D., and C. W. Lambiotte. 1971. Behavior of captive ornate box turtles. Journal of Biological Psychology 13:17–23.

Hart, D. R. 1983. Dietary and habitat shift with size of red-eared turtles (*Pseudemys scripta*) in a southern Louisiana population. Herpetologica 39:285–290.

Hay, R. W., and R. A. Christoffel. 2022. Ornate box turtle, *Terrapene ornata* (Agassiz 1857). Pp. 814–829 in J. M. Kapfer and D. J. Brown (eds.), Amphibians and Reptiles of Wisconsin. University of Wisconsin Press, Madison.

Hayes, F. E. 1989. Defensive stances in turtles. Herpetological Review 20:4–5.

Hedrick, A., and J. B. Iverson. 2017. *Kinosternon flavescens* (yellow mud turtle): Female growth and longevity. Herpetological Review 48:178–179.

Hill, E. C., H. J. Vogel, T. R. Rosburg, L. Lown, and K. S. Summerville. 2009. Response of a population of *Terrapene ornata ornata* to long distance relocation. Natural Areas Journal 29:311–316.

Hill, M. G. 2022. A Late Pleistocene wood turtle (*Glyptemys insculpta*) from Iowa, USA: Response of the taxon to glaciation and formation of the current range. Canadian Field-Naturalist 136:206–212.

House, W. J., I. M. Nall, and R. B. Thomas. 2010. Interpond movements of western painted turtles (*Chrysemys picta*) in east-central Kansas. Southwestern Naturalist 55:403–410.

Houseal, T. W., J. W. Bickham, and M. D. Springer. 1982. Geographic variation in the yellow mud turtle, *Kinosternon flavescens*. Copeia 1982:567–580.

Howing, D. T. L. 2021. Conservation Guidance for Blanding's Turtles in Northwest Iowa. Master's thesis, Montana State University, Bozeman.

Hughes, D. F., M. C. Allender, N. P. Bernstein, J. B. Iverson, C. Kolthoff, B. T. Martin, and B. M. Reed. In press. *Terrapene ornata* Agassiz (1857): Ornate box turtle, plains box turtle, western box turtle, desert box turtle, tortuga de caja ornamentada (Spanish), schmuck-dosenschildkröte (German). In A. G. J. Rhodin, J. B. Iverson, P. P. van Dijk, K. A. Buhlmann, P. C. H. Pritchard, and R. A. Mittermeier (eds.), Conservation Biology of Freshwater Turtles and Tortoises: A Compilation Project of the IUCN/SSC Tortoise and Freshwater Turtle Specialist Group. Chelonian Research Monographs.

Hughes, D. F., S. Dukuly, D. F. Becker, J. Scharosch, J. G. Otten, S. Wagner, and B. M. Reed. 2023. Copulation phenology of ornate box turltes (*Terrapene ornata*). Journal of Herpetoloty 57:246–253

Hurter, J. 1911. Herpetology of Missouri. Transactions of the Academy of Science of Saint Louis 20:59–274.

Hutchison, V. H., A. Vinegar, and R. J. Kosh. 1966. Critical thermal maxima in turtles. Herpetologica 22:32–41.

Iowa Department of Natural Resources. 2015. Securing a Future for Fish and Wildlife: A Conservation Legacy for Iowans. Iowa Department of Natural Resources, Des Moines.

Iverson, J. B. 1979. A taxonomic reappraisal of the yellow mud turtle, *Kinosternon flavescens* (Testudines: Kinosternidae). Copeia 1979:212–225.

Iverson, J. B. 1990. Nesting and parental care in the mud turtle, *Kinosternon flavescens*. Canadian Journal of Zoology 68:230–233.

Iverson, J. B. 1991. Life history and demography of the yellow mud turtle, *Kinosternon flavescens*. Herpetologica 47:373–395.

Iverson, J. B. 2022. Climate-mediated recruitment failure in a turtle population and its bearing on northern limits of distribution. Chelonian Conservation and Biology 21:181–186.

Iverson, J. B. 2023. Incidence of botfly infestation in the ornate box turtle (*Terrapene ornata*). Chelonian Conservation and Biology 22:230–232.

Iverson, J. B., and J. L. Christiansen. 2023. *Kinosternon flavescens* (Agassiz 1857): Yellow mud turtle, casquito amarillo. Pp. 121.1–121.26 in A. G. J. Rhodin, J. B. Iverson, P. P. van Dijk, C. B. Stanford, E. V. Goode, K. A. Buhlmann, and R. A. Mittermeier (eds.), Conservation Biology of Freshwater Turtles and Tortoises: A Compilation Project of the IUCN/SSC Tortoise and Freshwater Turtle Specialist Group. Chelonian Research Monographs 5.

Iverson, J. B., P. A. Meylan, and M. E. Seidel. 2017. Testudines—turtles. Pp. 1–102 in B. I. Crother (ed.), Scientific and Standard English Names of Amphibians and Reptiles of North America North of Mexico, with Comments Regarding Confidence in Our Understanding, 8th ed. SSAR Herpetological Circular 43.

Iverson, J. B., and G. R. Smith. 1993. Reproductive ecology of the painted turtle (*Chrysemys picta*) in the Nebraska Sandhills and across its range. Copeia 1993:1–21.

Jaeger, C. P., and V. A. Cobb. 2012. Comparative spatial ecologies of female painted turtles (*Chrysemys picta*) and red-eared sliders (*Trachemys scripta*) at Reelfoot Lake, Tennessee. Chelonian Conservation and Biology 11:59–67.

Janzen, F. J. 1993. The influence of incubation temperature and family on eggs, embryos, and hatchlings of the smooth softshell turtle (*Apalone mutica*). Physiological Zoology 66:349–373.

Janzen, F. J. 1994a. Climate change and temperature-dependent sex determination in reptiles. Proceedings of the National Academy of Sciences 91:7487–7490.

Janzen, F. J. 1994b. Vegetational cover predicts the sex ratio of hatchling turtles in natural nests. Ecology 75:1593–1599.

Janzen, F. J., and S. O'Steen. 1990. An instance of male combat in the common snapping turtle (*Chelydra serpentina*). Bulletin of the Chicago Herpetological Society 25:11.

Janzen, F. J., G. L. Paukstis, and E. D. Brodie. 1992. Observations on basking behavior of hatchling turtles in the wild. Journal of Herpetology 26:217–219.

Janzen, F. J., J. K. Tucker, and G. L. Paukstis. 2000a. Experimental analysis of an early life-history stage: Avian predation selects for larger body size of hatchling turtles. Journal of Evolutionary Biology 13:947–954.

Janzen, F. J., J. K. Tucker, and G. L. Paukstis. 2000b. Experimental analysis of an early life-history stage: Selection on size of hatchling turtles. Ecology 81:2290–2304.

Jenkins, J. D. 1979. Notes on the courtship of the map turtle, *Graptemys pseudogeographica* (Gray) (Reptilia, Testudines, Emydidae). Journal of Herpetology 13:129–131.

Johnson, T. R. 2000. The Amphibians and Reptiles of Missouri. 2nd ed. Missouri Department of Conservation, Jefferson City.

Jones, M. T., and L. L. Willey (eds.). 2021. Biology and Conservation of the Wood Turtle. Northeast Association of Fish and Wildlife Agencies, Petersburgh, N.Y.

Kaufmann, J. H. 1992a. Habitat use by wood turtles in central Pennsylvania. Journal of Herpetology 26:315–321.

Kaufmann, J. H. 1992b. The social behavior of wood turtles, *Clemmys insculpta*, in central Pennsylvania. Herpetological Monographs 6:1–25.

Klimstra, W. D. 1950. Notes on some amphibians and reptiles from Davis County, Iowa. Iowa State College Journal of Science 24:429–431.

Klimstra, W. D. 1951. Notes on late summer snapping turtle movements. Herpetologica 7:140.

Kofron, C. P., and A. A. Schreiber. 1985. Ecology of two endangered aquatic turtles in Missouri: *Kinosternon flavescens* and *Emydoidea blandingii*. Journal of Herpetology 19:27–40.

Kolbe, J. J., and F. J. Janzen. 2002a. Experimental analysis of an early life-history stage: Water loss and migrating hatchling turtles. Copeia 2002:220–226.

Kolbe, J. J., and F. J. Janzen. 2002b. Impact of nest-site selection on nest success and nest temperature in natural and disturbed habitats. Ecology 83:269–281.

Krawchuk, M. A., and R. J. Brooks. 1998. Basking behavior as a measure of reproductive cost and energy allocation in the painted turtle, *Chrysemys picta*. Herpetologica 54:112–121.

Kuehn, S. R. 2016. Zooarchaeological remains from the Crawford Farm Site (11R181): Nativist faunal exploitation and bone use by the historic Sauk in Illinois. Paper presented at the Midwest Archaeological Conference, University of Iowa, Iowa City.

Lagler, K. F. 1943. Food habits and economic relations of the turtles of Michigan with special reference to fish management. American Midland Naturalist 29:257–312.

Lambert, M. R., O. Hernández-Gómez, A. R. Krohn, A. Mutlow, L. Patterson, E. B. Rosenblum, M. Timmer, J. Willis, and J. Bushell. 2021. Turtle shell disease fungus (*Emydomyces testavorans*): First documented occurrence in California and prevalence in free-living turtles. Ichthyology and Herpetology 109:958–962.

Lamm, S. H., A. Taylor, Jr., E. J. Gangarosa, H. W. Anderson, W. Young, M. H. Clark, and A. R. Bruce. 1972. Turtle-associated salmonellosis: I. An estimation of the magnitude of the problem in the United States, 1970–1971. American Journal of Epidemiology 95:511–517.

Laporte, M., C. O. Silva Beaudry, and B. Angus. 2013. Effects of road proximity on genetic diversity and reproductive success of the painted turtle (*Chrysemys picta*). Conservation Genetics 14:21–30.

Lardie, R. L. 1975a. Courtship and mating behavior in the yellow mud turtle, *Kinosternon flavescens*. Journal of Herpetology 9:223–227.

Lardie, R. L. 1975b. Observations of reproduction in *Kinosternon*. Journal of Herpetology 9:260–264.

Lardie, R. L. 1978. Additional observations on courtship and mating in the plains yellow mud turtle, *Kinosternon flavescens flavescens*. Bulletin of the Oklahoma Herpetological Society 3:70–72.

Lardie, R. L. 1979. Eggs and young of the plains yellow mud turtle. Bulletin of the Oklahoma Herpetological Society 4:24–32.

Lardie, R. L. 1983. Aggressive interactions and territoriality in the yellow mud turtle, *Kinosternon flavescens flavescens* (Agassiz). Bulletin of the Oklahoma Herpetological Society 8:68–81.

LeClere, J. B. 2013. A Field Guide to the Amphibians and Reptiles of Iowa. ECO Herpetological Publishing, Rodeo, N.M.

Lefevre, K., and R. J. Brooks. 1995. Effects of sex and body size on basking behavior in a northern population of the painted turtle, *Chrysemys picta*. Herpetologica 51:217–224.

Legler, J. M. 1960. Natural history of the ornate box turtle, *Terrapene ornata ornata* Agassiz. University of Kansas Publications 11:527–669.

Lewis, E. L., and J. B. Iverson. 2018. *Terrapene ornata* (ornate box turtle) longevity. Herpetological Review 49:530.

Lewis, E. L., J. B. Iverson, G. R. Smith, and J. E. Rettig. 2018. Body size and growth in the red-eared slider (*Trachemys scripta elegans*) at the northern edge of its range: Does Bergmann's rule apply? Herpetological Conservation and Biology 13:700–710.

Linck, M., and J. J. Moriarty. 1998. Blanding's turtle use of a restored prairie and wetland complex in Crow-Hassan Reserve, Minnesota. Abstract, Turtle Workshop, May 7–8, J. F. Bell Museum of Natural History, Minneapolis.

Linck, M. H., J. A. DePari, B. O. Butler, and T. E. Graham. 1989. Nesting behavior of the turtle, *Emydoidea blandingi*, in Massachusetts. Journal of Herpetology 23:442–444.

Lindeman, P. V. 1999a. Aggressive interactions during basking among four species of emydid turtles. Journal of Herpetology 33:214–219.

Lindeman, P. V. 1999b. Surveys of basking map turtles *Graptemys* spp. in three river drainages and the importance of deadwood abundance. Biological Conservation 88:33–42.

Lindeman, P. V. 2001. A contrast in the basking habits of the sympatric trionychid turtles *Apalone mutica* and *A. spinifera*. Herpetological Natural History 8:87–89.

Lindeman, P. V. 2006. Zebra and quagga mussels (*Dreissena* spp.) and other prey of a Lake Erie population of northern map turtles (Emydidae: *Graptemys geographica*). Copeia 2006:268–273.

Lindeman, P. V. 2013. The Map Turtle and Sawback Atlas: Ecology, Evolution, Distribution, and Conservation. Vol. 12. University of Oklahoma Press, Norman.

Lindeman, P. V., and F. W. Rabe. 1990. Effect of drought on the western painted turtle, *Chrysemys picta belli*, in a small wetland ecosystem. Journal of Freshwater Ecology 5:359–364.

Long, D. R. 1985. Lipid utilization during reproduction in female *Kinosternon flavescens*. Herpetologica 41:58–65.

Long, D. R. 1986. Clutch formation in the turtle, *Kinosternon flavescens* (Testudines: Kinosternidae). Southwestern Naturalist 31:1–8.

Loomis, R. B. 1948. Notes on the herpetology of Adams County, Iowa. Herpetologica 4:121–122.

Lovich, J. E., C. J. McCoy, and W. R. Garstka. 1990. The development and significance of melanism in the slider turtle. Pp. 233–254 in J. W. Gibbons (ed.), Life History and Ecology of the Slider Turtle. Smithsonian Institution Press, Washington, D.C.

Luiselli, L., A. Starita, G. M. Carpaneto, G. H. Segniagbeto, and G. Amori. 2016. A short review of the international trade of wild tortoises and freshwater turtles across the world and throughout two decades. Chelonian Conservation and Biology 15:167–172.

Lundgren, S. A., K. Geluso, and C. W. Schoenebeck. 2012. Terrestrial and semi-aquatic vertebrates in diets of largemouth bass in central Nebraska. Prairie Naturalist 44:105–108.

Mahmoud, I. Y. 1967. Courtship behavior and sexual maturity in four species of kinosternid turtles. Copeia 1967:314–319.

Mali, I., M. W. Vandewege, S. K. Davis, and M. R. J. Forstner. 2014. Magnitude of the freshwater turtle exports from the US: Long term trends and early effects of newly implemented harvest management regimes. PLOS ONE 9:e86478.

Martin, B. T., N. P. Bernstein, R. D. Birkhead, J. F. Koukl, S. M. Mussmann, and J. S. Placyk, Jr. 2013. Sequence-based molecular phylogenetics and phylogeography of the American box turtles (*Terrapene* spp.) with support from DNA barcoding. Molecular Phylogenetics and Evolution 68:119–134.

Martin, B. T., N. P. Bernstein, R. D. Birkhead, J. F. Koukl, S. M. Mussmann, and J. S. Placyk, Jr. 2014. On the reclassification of the *Terrapene* (Testudines: Emydidae): A response to Fritz and Havaš. Zootaxa 3835:292–294.

Martin, B. T., M. R. Douglas, T. K. Chafin, J. S. Placyk, Jr., R. D. Birkhead, C. A. Phillips, and M. E. Douglas. 2020. Contrasting signatures of introgression in North American box turtle (*Terrapene* spp.) contact zones. Molecular Ecology 29:4186–4202.

McCallum, M. L. 2021. Turtle biodiversity losses suggest coming sixth mass extinction. Biodiversity and Conservation 30:1257–1275.

Meeks, R. L., and G. R. Ultsch. 1990. Overwintering behavior of snapping turtles. Copeia 1990:880–884.

Mitchell, S. M., L. K. Muehlbauer, and S. Freedberg. 2016. Nuclear introgression without mitochondrial introgression in two turtle species exhibiting sex-specific trophic differentiation. Ecology and Evolution 6:3280–3288.

Moll, D. 1976. Food and feeding strategies of the Ouachita map turtle (*Graptemys pseudogeographica ouachitensis*). American Midland Naturalist 96:478–482.

Moll, D. L. 1974. Notes on the behavior of Isle Royale painted turtles (*Chrysemys picta bellii*). Journal of Herpetology 8:254–255.

Moll, D. L. 1979. Subterranean feeding by the Illinois mud turtle, *Kinosternon flavescens spooneri*. Journal of Herpetology 13:371–373.

Moll, E. O., and D. Moll. 2000. Conservation of river turtles. Pp. 126–155 in M. W. Klemens (ed.), Turtle Conservation. Smithsonian Institution Press, Washington, D. C.

Moorman, R. B., and K. D. Carlander. 1951. Snapper and soft-shells. Iowa Conservationist 10:129, 135.

Moriarty, J. J., and C. D. Hall. 2014. Amphibians and Reptiles in Minnesota. University of Minnesota Press, Minneapolis.

Morrison, M. B., B. P. Butterfield, S. G. Ross, C. Collins, A. Walde, J. Gray, J. B. Hauge, and E. C. Munscher. 2019. The diet of the eastern musk turtle (*Sternotherus odoratus*) as it pertains to invasive snail consumption in a freshwater spring habitat in Texas. Herpetology Notes 12:1133–1139.

Muller, J. F. 1921. Notes on the habits of the soft-shell turtle, *Amyda mutica*. American Midland Naturalist 7:180–184.

Murphy, J. C., and M. J. Corn. 1977. A turtle vanishes. Letter. Natural History 86:8.

Nagle, R. D., C. L. Lutz, and A. L. Pyle. 2004. Overwintering in the nest by hatchling map turtles (*Graptemys geographica*). Canadian Journal of Zoology 82:1211–1218.

Neuman-Lee, L. A., and F. J. Janzen. 2011. Atrazine exposure impacts behavior and survivorship of neonatal turtles. Herpetologica 67:23–31.

Newman, H. H. 1906. The habits of certain tortoises. Journal of Comparative Neurology and Psychology 16:126–152.

Noble, G. K., and A. M Breslau. 1938. The senses involved in the migration of young fresh-water turtles after hatching. Journal of Comparative Psychology 25:175–193.

Nutting, W. L., and T. E. Graham. 1993. Preferred body temperatures in five nearctic freshwater turtles: A preliminary study. Comparative Biochemistry and Physiology Part A: Physiology 104:243–246.

Obbard, M. E., and R. J. Brooks. 1987. Prediction of the onset of the annual nesting season of the common snapping turtle, *Chelydra serpentina*. Herpetologica 43:324–328.

Olson, D. H., and D. Saenz. 2013. Climate Change and Reptiles. U.S. Department of Agriculture, Forest Service, Climate Change Resource Center. Available at https://www.fs.usda.gov/ccrc/topics/reptiles-and-climate-change.

Osborn, H. 1892. A partial catalogue of the animals of Iowa, represented in the collections of the Department of Zoology and Entomology of the Iowa Agricultural College (1891), Ames. 1–39.

Otten, J. G. 2017. Factors Influencing Wood Turtle (*Glyptemys insculpta*) Home Range Size in Iowa: A Comparison between Suburban and Rural Populations. Master's thesis, University of Northern Iowa, Cedar Falls.

Otten, J. G. 2020. Twenty-four new county and river drainage records for *Graptemys* spp. for Iowa, USA. Herpetological Review 51:284–286.

Otten, J. G., I. T. Clifton, D. F. Becker, and J. M. Refsnider. 2024. Here comes the sun: Thermoregulatory behavior in ectotherms illuminated by light-level geolocators. Journal of Zoology 322, available at https://doi.org/10.1111/jzo.13168.

Otten, J. G., A. C. Hulbert, S. W. Berg, and J. W. Tamplin. 2021. Home range, site fidelity, and movement patterns of the wood turtle (*Glyptemys insculpta*) at the southwestern edge of its range. Chelonian Conservation and Biology 20:231–241.

Otten, J. G., and T. J. VanDeWalle. 2014. *Apalone spinifera* (spiny softshell), *Chelydra serpentina* (snapping turtle), and *Sternotherus odoratus* (eastern musk turtle): Homing behavior. Herpetological Review 45:309–311.

Padgett, D. J., J. J. Carboni, and D. J. Schepis. 2010. The dietary composition of *Chrysemys picta picta* (eastern painted turtles) with special reference to the seeds of aquatic macrophytes. Northeastern Naturalist 17:305–312.

Paisley, R. N., and R. W. Hay. 2022. Smooth softshell, *Apalone mutica* (LeSueur 1827). Pp. 840–849 in J. M. Kapfer and D. J. Brown (eds.), Amphibians and Reptiles of Wisconsin. University of Wisconsin Press, Madison.

Palmer, E. L. 1924. A note on the distribution of the wood turtle *Clemmys insculpta* (Le Conte). Copeia 1924:40.

Pappas, M. J., and B. J. Brecke. 1992. Habitat selection of juvenile Blanding's turtles, *Emydoidea blandingii*. Journal of Herpetology 26:233–234.

Pappas, M. J., B. J. Brecke, and J. D. Congdon. 2000. The Blanding's turtles (*Emydoidea blandingii*) of Weaver Dunes, Minnesota. Chelonian Conservation and Biology 3:557–568.

Parmenter, R. R., and H. W. Avery. 1990. The feeding ecology of the slider turtle. Pp. 257–265 in J. W. Gibbons (ed.), Life History and Ecology of the Slider Turtle. Smithsonian Institution Press, Washington, D.C.

Patterson, J. C., and P. V. Lindeman. 2009. Effects of zebra and quagga mussel (*Dreissena* spp.) invasion on the feeding habits of *Sternotherus odoratus* (stinkpot) on Presque Isle, northwestern Pennsylvania. Northeastern Naturalist 16:365–374.

Pearse, D. E., F. J. Janzen, and J. C. Avise. 2002. Multiple paternity, sperm storage, and reproductive success of female and male painted turtles (*Chrysemys picta*) in nature. Behavioral Ecology and Sociobiology 51:164–171.

Penn, G. H. 1950. Utilization of crawfishes by cold-blooded vertebrates in the eastern United States. American Midland Naturalist 44:643–658.

Phillips, C. A., R. A. Brandon, and E. O. Moll. 1999. Field Guide to Amphibians and Reptiles of Illinois. Illinois Natural History Survey Manual 8, Champaign.

Picard, G., M. A. Carrière, and G. Blouin-Demers. 2011. Common musk turtles (*Sternotherus odoratus*) select habitats of high thermal quality at the northern extreme of their range. Amphibia-Reptilia 32:83–92.

Piepgras, S. A., and J. W. Lang. 2000. Spatial ecology of Blanding's turtle in central Minnesota. Chelonian Conservation and Biology 3:589–601.

Platt, W. J. 1973. Comparison of vertebrate communities of Coralville Reservoir and Cone Marsh, Iowa. Proceedings of the Iowa Academy of Science 80:117–128.

Platt, W. J. 1975. The vertebrate fauna of Cayler Prairie Preserve, Dickinson County, Iowa. Proceedings of the Iowa Academy of Science 82:106–108.

Plummer, M. V. 1977a. Activity, habitat and population structure in the turtle, *Trionyx muticus*. Copeia 1977:431–440.

Plummer, M. V. 1977b. Notes on the courtship and mating behavior of the soft-shell turtle, *Trionyx muticus* (Reptilia, Testudines, Trionychidae). Journal of Herpetology 11:90–92.

Plummer, M. V. 1977c. Reproduction and growth in the turtle *Trionyx muticus*. Copeia 1977:440–447.

Plummer, M. V., and J. C. Burnley. 1997. Behavior, hibernacula, and thermal relations of softshell turtles (*Trionyx spiniferus*) overwintering in a small stream. Chelonian Conservation and Biology 2:489–493.

Plummer, M .V., T. L. Crabill, N. E. Mills, and S. L. Allen. 2005. Body temperatures of free-ranging softshell turtles (*Apalone spinifera*) in a small stream. Herpetological Review 36:371–374.

Plummer, M. V., and D. B. Farrar. 1981. Sexual dietary differences in a population of *Trionyx muticus*. Journal of Herpetology 15:175–179.

Plummer, M. V., and J. W. Goy. 1997. *Sternotherus odoratus* (common musk turtle): Mortality. Herpetological Review 28:88.

Plummer, M. V., T. N. Lee, and N. E. Mills. 2008. Effect of sand substrate on the growth and condition of *Apalone mutica* hatchlings. Journal of Herpetology 42:550–554.

Plummer, M. V., N. E. Mills, and S. L. Allen. 1997. Activity, habitat, and movement patterns of softshell turtles (*Trionyx spiniferus*) in a small stream. Chelonian Conservation and Biology 2:514–520.

Plummer, M. V., C. E. Shadrix, and R. C. Cox. 1994. Thermal limits of incubation in embryos of softshell turtles (*Apalone mutica*). Chelonian Conservation and Biology 1:141–144.

Plummer, M. V., and H. W. Shirer. 1975. Movement patterns in a river population of the softshell turtle, *Trionyx muticus*. University of Kansas Museum of Natural History Occasional Papers 43:1–26.

Pluto, T. G., and E. D. Bellis. 1986. Habitat utilization by the turtle, *Graptemys geographica*, along a river. Journal of Herpetology 20:22–31.

Pluto, T. G., and E. D. Bellis. 1988. Seasonal and annual movements of riverine map turtles, *Graptemys geographica*. Journal of Herpetology 22:152–158.

Pope, C. H. 1939. Turtles of the United States and Canada. A. A. Knopf, New York.

Powell, M. A., and R. Powell. 2011. Aquatic turtles feasting on periodical cicadas. Missouri Herpetological Association Newsletter 24:21.

Powell, R., R. Conant, and J. T. Collins. 2016. Peterson Field Guide to Reptiles and Amphibians of Eastern and Central North America. 4th ed. Peterson Field Guide Series. Houghton Mifflin Harcourt, Boston.

Prior, J. C. 1991. Landforms of Iowa. University of Iowa Press, Iowa City.

Quinn, A. J., and J. L. Christiansen. 1972. The relationship between pond bottom type and growth rate of western painted turtles *Chrysemys picta belli* in Iowa, a preliminary report. Proceedings of the Iowa Academy of Science 78:67–69.

Ratterman, R. J., and R. A. Ackerman. 1989. The water exchange and hydric microclimate of painted turtle (*Chrysemys picta*) eggs incubating in field nests. Physiological Zoology 62:1059–1079.

Redder, A. J., C. K. Dodd, Jr., D. Keinath, D. McDonald, and T. Ise. 2006. Ornate box turtle (*Terrapene ornata ornata*): A Technical Conservation Assessment. USDA Forest Service, Rocky Mountain Region, U.S. Department of Agriculture. Available at http://www.fs.fed.us/r2/projects/scp/assessments/ornateboxturtle.pdf.

Reed, B. M., K. Hobelman, A. Gauntt, M. Schwenka, A. Trautman, P. Wagner, S. Kim, C. Armstrong, S. Wagner, A. Weller, K. Brighton, S. Bloom, C. Nelson, F. Suboh, C. Kolthoff, S. Dukuly, R. J. Mercader, and D. F. Hughes. 2023. Spatiotemporal variation of behavior and repeatability in a long-lived turtle. Behavioral Ecology and Sociobiology 77:88.

Reese, S. A., D. C. Jackson, and G. R. Ultsch. 2003. Hibernation in freshwater turtles: Softshell turtles (*Apalone spinifera*) are the most intolerant of anoxia among North American species. Journal of Comparative Physiology B 173:263–268.

Richards-Dimitrie, T., S. E. Gresens, S. A. Smith, and R. A. Seigel. 2013. Diet of northern map turtles (*Graptemys geographica*): Sexual differences and potential impacts of an altered river system. Copeia 2013:477–484.

Richtsmeier, R. J. 2005. Migration, Gene Flow, and Genetic Diversity within and among Populations of Ornate Box Turtles (*Terrapene ornata ornata*). Master's thesis, University of Northern Iowa, Cedar Falls.

Richtsmeier, R. J., N. P. Bernstein, J. W. Demastes, and R. W. Black. 2008. Migration, gene flow, and genetic diversity within and among Iowa populations of ornate box turtles (*Terrapene ornata ornata*). Chelonian Conservation and Biology 7:3–11.

Rizkalla, C. E., and R. K. Swihart. 2006. Community structure and differential responses of aquatic turtles to agriculturally induced habitat fragmentation. Landscape Ecology 21:1361–1375.

Robinson, K. M., and G. G. Murphy. 1978. The reproductive cycle of the eastern spiny softshell turtle (*Trionyx spiniferus spiniferus*). Herpetologica 34:137–140.

Rollinson, N., G. J. Tattersall, and R. J. Brooks. 2008. Overwintering habitats of a northern population of painted turtles (*Chrysemys picta*): Winter temperature selection and dissolved oxygen concentrations. Journal of Herpetology 42:312–321.

Root, R. W. 1949. Aquatic respiration in the musk turtle. Physiological Zoology 22:172–178.

Rose, F. L. 1980. Turtles in arid and semi-arid regions. Bulletin of the Ecological Society of America 61:89.

Ross, D. A., and R. K. Anderson. 1990. Habitat use, movements, and nesting of *Emydoidea blandingi* in central Wisconsin. Journal of Herpetology 24:6–12.

Ross, D. A., and J. E. Lovich. 1992. Does the color pattern of two species of turtles imitate duckweed? Journal of the Pennsylvania Academy of Science 66:39–42.

Rowe, J. W. 1987. Seasonal and Daily Activity in a Population of Blanding's Turtle (*Emydoidea blandingi*) in Northern Illinois. Master's thesis, Eastern Illinois University, Charleston.

Rowe, J. W. 1992a. Dietary habits of the Blanding's turtle (*Emydoidea blandingi*) in northeastern Illinois. Journal of Herpetology 26:111–114.

Rowe, J. W. 1992b. Observations of body size, growth, and reproduction in Blanding's turtle (*Emydoidea blandingii*) from western Nebraska. Canadian Journal of Zoology 70:1690–1695.

Rowe, J. W., and S. F. Dalgarn. 2010. Home range size and daily movements of midland painted turtles (*Chrysemys picta marginata*) in relation to body size, sex, and weather patterns. Herpetological Conservation and Biology 5:461–473.

Ruthven, A. G. 1910. Contributions to the herpetology of Iowa. Proceedings of the Iowa Academy of Science 17:198–209.

Ryan, K. M., and P. V. Lindeman. 2007. Reproductive allometry in the common map turtle, *Graptemys geographica*. American Midland Naturalist 158:49–59.

Ryan, T. J., C. A. Conner, B. A. Douthitt, S. C. Sterrett, and C. M. Salsbury. 2008. Movement and habitat use of two aquatic turtles (*Graptemys geographica* and *Trachemys scripta*) in an urban landscape. Urban Ecosystems 11:213–225.

Sajwaj, T. D., and J. W. Lang. 2000. Thermal ecology of Blanding's turtle in central Minnesota. Chelonian Conservation and Biology 3:626–636.

Schueler, F. W. 1983. Reticulate melanism in Canadian western painted turtles. Blue Jay 41:83–91.

Scott, T. G. 1938. Wildlife mortality on Iowa highways. American Midland Naturalist 20:527–539.

Seidel, M. E. 1978. Terrestrial dormancy in the turtle *Kinosternon flavescens*: respiratory metabolism and dehydration. Comparative Biochemistry and Physiology Part A: Physiology 61:1–4.

Seidel, M. E., and S. L. Reynolds. 1980. Aspects of evaporative water loss in the mud turtles *Kinosternon hirtipes* and *Kinosternon flavescens*. Comparative Biochemistry and Physiology Part A: Physiology 67:593–598.

Semlitsch, R. D., and J. W. Gibbons. 1989. Lack of largemouth bass predation of hatchling turtles (*Trachemys scripta*). Copeia 1989:1030–1031.

Sethuraman, A., S. E. McGaugh, M. L. Becker, C. H. Chandler, J. L. Christiansen, S. Hayden, A. LeClere, J. Monson-Miller, E. M. Myers, R. T. Paitz, J. M. Refsnider, T. J. VanDeWalle, and F. J. Janzen. 2013. Population genetics of Blanding's turtle (*Emys blandingii*) in the midwestern United States. Conservation Genetics 2013:1–13.

Sexton, O. J. 1959. Spatial and temporal movements of a population of the painted turtle, *Chrysemys picta marginata* (Agassiz). Ecological Monographs 29:113–140.

Shane, S. M., R. Gilbert, and K. S. Harrington. 1990. Salmonella colonization in commercial pet turtles (*Pseudemys scripta elegans*). Epidemiology and Infection 105:307–316.

Smith, E. N., S. L. Robertson, and S. R. Adams. 1981. Thermoregulation of the spiny soft-shelled turtle *Trionyx spinifer*. Physiological Zoology 54:74–80.

Smith, G. R., and J. B. Iverson. 2004. Diel activity patterns of the turtle assemblage of a northern Indiana lake. American Midland Naturalist 152:156–164.

Smith, G. R., J. B. Iverson, and J. E. Rettig. 2006. Changes in a turtle community from a northern Indiana lake: A long-term study. Journal of Herpetology 40:180–185.

Smith, P. W. 1951. A new frog and a new turtle from the western Illinois sand prairies. Bulletin of the Chicago Academy of Sciences 9:189–199.

Smith, P. W. 1961. The Amphibians and Reptiles of Illinois. Illinois Natural History Survey Bulletin 28, Champaign.

Snider, A. T., and J. K. Bowler. 1992. Longevity of Reptiles and Amphibians in North American Collections. 2nd ed. Society for the Study of Amphibians and Reptiles, Ithaca, N.Y.

Somes, M. P. 1911. Notes on some Iowa reptiles. Proceedings of the Iowa Academy of Science 18:149–154.

Speaker, E. B. 1942. Mi-She-ke, the turtle, a very tasty fellow. Iowa Conservationist 1(10):5–6.

Spencer, R. J., and F. J. Janzen. 2011. Hatching behavior in turtles. Integrative and Comparative Biology 51:100–110.

Spradling, T. A., J. W. Tamplin, S. S. Dow, and K. J. Meyer. 2010. Conservation genetics of a peripherally isolated population of the wood turtle (*Glyptemys insculpta*) in Iowa. Conservation Genetics 11:1667–1677.

Steffen, J. E., K. M. Learn, J. S. Drumheller, S. M. Boback, and K. J. McGraw. 2015. Carotenoid composition of colorful body stripes and patches in the painted turtle (*Chrysemys picta*) and red-eared slider (*Trachemys scripta*). Chelonian Conservation and Biology 14:56–63.

Stone, P. A., J. L. Dobie, and R. P. Henry. 1992. Cutaneous surface area and bimodal respiration in soft-shelled (*Trionyx spiniferus*), stinkpot (*Sternotherus odoratus*), and mud turtles (*Kinosternon subrubrum*). Physiological Zoology 65:311–330.

Stoner, D. 1936. Wildlife casualties on the highways. Wilson Bulletin 48:276–283.

Storey, K. B. 2006. Reptile freeze tolerance: Metabolism and gene expression. Cryobiology 52:1–16.

Stromsten, F. A. 1923. Nest digging and egg laying habits of Bell's turtle *Crysemys marginata bellii* (Gray). University of Iowa Studies in Natural History 10:67–70.

Sutton, K. K. 1994. The Distribution and Ecology of Two Kinosternid Turtles in Iowa. Master's thesis, Drake University, Des Moines.

Sutton, K. K., and J. L. Christiansen. 1999. The habitat and distribution of the stinkpot, *Sternotherus odoratus*, in Iowa. Journal of the Iowa Academy of Science 106:63–65.

Tamplin, J. W. 2006. *Glyptemys insculpta* (wood turtle): Diet. Herpetological Review 37:456–457.

Tamplin, J. W. 2009. Effect of age and body size on selected temperature by juvenile wood turtles (*Glyptemys insculpta*). Journal of Thermal Biology 34:41–48.

Tamplin, J. W. 2010. *Apalone spinifera* (spiny softshell turtle): Maximum clutch size. Herpetological Review 41:344–345.

Tamplin, J. W. 2019. Upper Midwest turtle conservation—phase 2. Final Report to the U.S. Fish and Wildlife Service for Iowa State Wildlife Grant (No. F17AP00028).

Tamplin, J. W., J. B. Snodgrass, and D. B. Yeoman. 2009. *Glyptemys insculpta* (wood turtle): Diet. Herpetological Review 40:74–75.

Taylor, G. M., and E. Nol. 1989. Movements and hibernation sites of overwintering painted turtles in southern Ontario. Canadian Journal of Zoology 67:1877–1881.

Temple, S. A. 1987. Predation on turtle nests increases near ecological edges. Copeia 1987:250–252.

Thomas, E. S., and M. B. Trautman. 1937. Segregated hibernation of *Sternotherus odoratus* (Latreille). Copeia 1937:231.

Thorbjarnarson, J., C. J. Lagueux, D. Bolze, M. W. Klemens, and A. B. Meylan. 2000. Human use of turtles: A worldwide perspective. Pp. 33–84 in M. W. Klemens (ed.), Turtle Conservation. Smithsonian Institution Press, Washington, D.C.

Timken, R. L. 1968. *Graptemys pseudogeographica* in the upper Missouri River of the northcentral United States. Journal of Herpetology 1:76–82.

Tinkle, D. W. 1961. Geographic variation in reproduction, size, sex ratio and maturity of *Sternothaerus odoratus* (Testudinata: Chelydridae). Ecology 42:68–76.

Tinklepaugh, O. L. 1932. Maze learning of a turtle. Journal of Comparative Psychology 13:201.

Tucker, C. R., J. T. Strickland, B. S. Edmond, D. K. Delaney, and D. B. Ligon. 2015. Activity patterns of ornate box turtles (*Terrapene ornata*) in northwestern Illinois. Copeia 2015:502–511.

Tucker, J. K., C. R. Dolan, J. T. Lamer, and E. A. Dustman. 2008. Climatic warming, sex ratios, and red-eared sliders (*Trachemys scripta elegans*) in Illinois. Chelonian Conservation and Biology 7:60–69.

Tucker, J. K., S. A. Gritters, and R. A. Hrabik. 2008. Turtle Communities in the Upper Mississippi River System, 1992–1995. Illinois Natural History Survey Technical Report 30, Champaign.

Tucker, J. K., G. L. Paukstis, and F. J. Janzen. 1998. Annual and local variation in reproduction in the red-eared slider, *Trachemys scripta elegans*. Journal of Herpetology 32:515–526.

Tucker, J. K., G. L. Paukstis, and F. J. Janzen. 2001. Interspecific variation in egg and hatchling size in four sympatric chelonians: Influence of the hydric environment on incubation. Bulletin of the Maryland Herpetological Society 37:83–100.

Tuma, M. W. 1993a. Ecology of the State Endangered Yellow Mud Turtle, *Kinosternon flavescens*, in Henry Co., Illinois. Master's thesis, Eastern Illinois University, Charleston.

Tuma, M. W. 1993b. *Kinosternon flavescens* (yellow mud turtle): Multiple nesting. Herpetological Review 24:31.

Tuma, M. W. 2006. Range, habitat use, and seasonal activity of the yellow mud turtle (*Kinosternon flavescens*) in northwestern Illinois: Implications for site-specific conservation and management. Chelonian Conservation and Biology 5:108–120.

Turtle Taxonomy Working Group (TTWG) [A. G. J. Rhodin, J. B. Iverson, R. Bour, U. Fritz, A. Georges, H. B. Shaffer, and P. P. van Dijk.] 2021. Turtles of the World:

Annotated Checklist and Atlas of Taxonomy, Synonymy, Distribution, and Conservation Status. 9th ed. Chelonian Research Monographs.

Ultsch, G. R. 1985. The viability of nearctic freshwater turtles submerged in anoxia and normoxia at 3 and 10 degrees C. Comparative Biochemistry and Physiology Part A: Physiology 81:607–611.

Ultsch, G. R. 2006. The ecology of overwintering among turtles: Where turtles overwinter and its consequences. Biological Reviews 81:339–367.

Ultsch, G. R., M. E. Carwile, C. E. Crocker, and D. C. Jackson. 1999. The physiology of hibernation among painted turtles: The eastern painted turtle *Chrysemys picta picta*. Physiological Biochemistry and Physiology 72:493–501.

Ultsch, G. R., C. V. Herbert, and D. C. Jackson. 1984. The comparative physiology of diving in North American freshwater turtles, part 1: Submergence tolerance, gas exchange, and acid-base balance. Physiological Zoology 57:620–631.

Ultsch, G. R., and D. Lee. 1983. Radiotelemetric observations of wintering snapping turtles (*Chelydra serpentina*) in Rhode Island. Journal of the Alabama Academy of Science 54:200–206.

VanDeWalle, T. 2011. Turtles in your pocket: A guide to freshwater and terrestrial turtles of the Upper Midwest. University of Iowa Press, Iowa City.

VanDeWalle, T. J. 1993. The Impact of Stream Modification on the Distribution of Riverine Turtles in Iowa. Master's thesis, Drake University, Des Moines.

VanDeWalle, T. J. 2021. *Emydoidea blandingii* (Blanding's turtle): Mating. Herpetological Review 52:846.

VanDeWalle, T. J., and J. L. Christiansen. 1996. A relationship between river modification and species richness of freshwater turtles in Iowa. Journal of the Iowa Academy of Science 103:1–8.

VanDeWalle, T. J., and J. G. Otten. 2014. Habitat selection and survivorship of hatchling Blanding's turtles (*Emydoidea blandingii*) in Iowa. Unpublished report submitted to the Iowa Department of Natural Resources, Des Moines.

VanDeWalle, T. J., and J. G. Otten. 2022a. *Emydoidea blandingii* (Blanding's turtle): Fertility and hatchling survival. Herpetological Review 53:666–667.

VanDeWalle, T. J., and J. G. Otten. 2022b. *Emydoidea blandingii* (Blanding's turtle): Terrestrial estivation. Herpetological Review 53:667–668.

VanDeWalle, T. J., and S. J. Parks. 2021. *Emydoidea blandingii* (Blanding's turtle): Hibernation. Herpetological Review 52:845–846.

Vogt, R. C. 1978. Systematics and Ecology of the False Map Turtle Complex *Graptemys pseudogeographica*. Ph.D. dissertation, University of Wisconsin–Madison.

Vogt, R. C. 1979. Cleaning/feeding symbiosis between grackles (*Quiscalus*: Icteridae) and map turtles (*Graptemys*: Emydidae). Auk 96:608–609.

Vogt, R. C. 1980. Natural history of the map turtles *Graptemys pseudogeographica* and *G. ouachitensis* in Wisconsin. Tulane Studies in Zoology and Botany 22:17–48.

Vogt, R. C. 1981a. Food partitioning among three species of *Graptemys*. American Midland Naturalist 105:102–111.

Vogt, R. C. 1981b. Natural History of Amphibians and Reptiles of Wisconsin. Milwaukee Public Museum, Milwaukee.

Vogt, R. C. 1981c. Turtle egg (*Graptemys*: Emydidae) infestation by fly larvae. Copeia 1981:457–459.

Vogt, R. C. 1993. Systematics of the false map turtles (*Graptemys pseudogeographica* complex: Reptilia, Testudines, Emydidae). Annals of the Carnegie Museum 62:1–46.

Vogt, R. C. 2018. *Graptemys ouachitensis* Cagle 1953: Ouachita map turtle. Pp. 103.1–13 in A. G. J. Rhodin, J. B. Iverson, P. P. van Dijk, K. A. Buhlmann, P. C. H. Pritchard, and R. A. Mittermeier (eds.), Conservation Biology of Freshwater Turtles and Tortoises: A Compilation Project of the IUCN/SSC Tortoise and Freshwater Turtle Specialist Group. Chelonian Research Monographs 5.

Vogt, R. C. 2022a. False map turtle, *Graptemys pseudogeographica* (Gray 1831). Pp. 804–813 in J. M. Kapfer and D. J. Brown (eds.), Amphibians and Reptiles of Wisconsin. University of Wisconsin Press, Madison.

Vogt, R. C. 2022b. Northern map turtle, *Graptemys geographica* (Le Sueur 1817). Pp. 784–793 in J. M. Kapfer and D. J. Brown (eds.), Amphibians and Reptiles of Wisconsin. University of Wisconsin Press, Madison.

Vogt, R. C. 2022c. Ouachita map turtle, *Graptemys ouachitensis* (Cagle 1953). Pp. 794–803 in J. M. Kapfer and D. J. Brown (eds.), Amphibians and Reptiles of Wisconsin. University of Wisconsin Press, Madison.

Vogt, R. C., and J. J. Bull. 1982. Temperature-controlled sex determination in turtles: Ecological and behavioral aspects. Herpetologica 38:156–164.

Vogt, R. C., and J. J. Bull. 1984. Ecology of hatchling sex ratio in map turtles. Ecology 65:582–587.

Vogt, R. C., J. J. Bull, C. J. McCoy, and T. W. Houseal. 1982. Incubation temperature influences sex determination in kinosternid turtles. Copeia 1982:480–482.

Vogt, R. C., G. Bulté, and J. B. Iverson. 2018. *Graptemys geographica* (LeSueur 1817): Northern map turtle, common map turtle. Pp. 104.1–18 in A. G. J. Rhodin, J. B. Iverson, P. P. van Dijk, K. A. Buhlmann, P. C. H. Pritchard, and R. A. Mittermeier (eds.), Conservation Biology of Freshwater Turtles and Tortoises: A Compilation Project of the IUCN/SSC Tortoise and Freshwater Turtle Specialist Group. Chelonian Research Monographs 5.

Wacha, R. S., and J. L. Christiansen. 1976. Coccidian parasites from Iowa turtles: Systematics and prevalence. Journal of Protozoology 23:57–63.

Wacha, R. S., and J. L. Christiansen. 1977. Additional notes on the coccidian parasites of the soft-shell turtle, *Trionyx spiniferus* Le Sueur, in Iowa, with a description of *Eimeria vesicostieda* sp. n. Journal of Protozoology 24:357–359.

Wacha, R. S., and J. L. Christiansen. 1979. *Eimeria filamentifera* sp. n. from the snapping turtle, *Chelydra serpentina* (Linné), in Iowa. Journal of Protozoology 26:353–354.

Walde, A. D., J. R. Bider, D. Masse, R. A. Saumure, and R. D. Titman. 2007. Nesting ecology and hatching success of the wood turtle, *Glyptemys insculpta*, in Quebec. Herpetological Conservation and Biology 2:49–60.

Washburn, L. 2007. Death on the dunes. Iowa Outdoors March/April:47–53.

Watermolen, D. J. 2004. Softshell turtles (*Apalone* spp.) as bald eagle prey. Bulletin of the Chicago Herpetological Society 39:69–70.

Watts, M. R., and J. L. Christiansen. 1989. Turtle use and succession in a pool created in the bottom of a dry lake. Abstracts of Papers, 101st Session, Iowa Academy of Science, April 21–22, Buena Vista College. Journal of the Iowa Academy of Science 96:A37.

Webb, R. G. 1961. Observations on the life histories of turtles (genus *Pseudemys* and *Graptemys*) in Lake Texoma, Oklahoma. American Midland Naturalist 65:193–214.

Webb, R. G. 1962. North American recent soft-shelled turtles (family Trionychidae). University of Kansas Museum of Natural History 13:429–611.

White, D., and D. Moll. 1991. Clutch size and annual reproductive potential of the turtle *Graptemys geographica* in a Missouri stream. Journal of Herpetology 25:493–494.

White, D., and D. Moll. 1992. Restricted diet of the common map turtle *Graptemys geographica* in a Missouri stream. Southwestern Naturalist 37:317–318.

White, J. B., and G. G. Murphy. 1973. The reproductive cycle and sexual dimorphism of the common snapping turtle, *Chelydra serpentina serpentina*. Herpetologica 29:240–246.

Williams, J. 2013. Habitat Usage, Movement Patterns, and Home Range Size of Wood Turtles (*Glyptemys insculpta*) in a Suburban Habitat. Master's thesis, University of Northern Iowa, Cedar Falls.

Williams, J. E. 1952. Homing behavior of the painted turtle and musk turtle in a lake. Copeia 1952:76–82.

Williams, T. A., and J. L. Christiansen. 1981. The niches of two sympatric softshell turtles, *Trionyx muticus* and *Trionyx spiniferus*, in Iowa. Journal of Herpetology 15:303–308.

Wilsey, C., B. Bateman, L. Taylor, J. X. Wu, G. LeBaron, R. Shepherd, C. Koseff, S. Friedman, and R. Stone. 2019. Survival by Degrees: 389 Bird Species on the Brink. National Audubon Society, New York.

Yarrow, H. C. 1882. Check-list of North American reptilia and batrachia, with catalogue of specimens in the United States National Museum. U.S. National Museum Bulletin 24:1–249.

Yeomans, R. S. 1995. Water-finding in adult turtles: Random search or oriented behaviour? Animal Behaviour 49:977–987.

LIFE LIST

Use this section to keep a life list of the Iowa turtles you see and identify. Check the species you see, note the date you see them, and make other notes about them. In this way, this book can become your personal lifetime natural history diary of each new sighting.

- ☐ Blanding's Turtle
- ☐ Eastern Musk Turtle
- ☐ Eastern Spiny Softshell
- ☐ Midland Smooth Softshell
- ☐ Northern False Map Turtle
- ☐ Northern Map Turtle
- ☐ Ornate Box Turtle
- ☐ Ouachita Map Turtle
- ☐ Red-Eared Slider
- ☐ Snapping Turtle
- ☐ Western Painted Turtle
- ☐ Wood Turtle
- ☐ Yellow Mud Turtle

Photo by Wendy VanDeWalle.

THE END.

INDEX OF SCIENTIFIC AND COMMON NAMES

Boldface numbers indicate detailed discussion.

BUR OAK GUIDES

Field Guide to Wildflowers of Nebraska and the Great Plains: Second Edition
by Jon Farrar

Fish in Your Pocket: A Guide to Fish of the Upper Midwest
by Terry VanDeWalle

Forest and Shade Trees of Iowa: Third Edition
by Peter J. van der Linden and Donald R. Farrar

Frogs and Toads in Your Pocket: A Guide to Amphibians of the Upper Midwest
by Terry VanDeWalle

Mushrooms and Other Fungi of the Midcontinental United States: Second Edition
by Donald M. Huffman, Lois H. Tiffany, George Knaphus, and Rosanne A. Healy

The Natural History of the Snakes and Lizards of Iowa
by Terry VanDeWalle

The Prairie in Seed: Identifying Seed-Bearing Prairie Plants in the Upper Midwest
by Dave Williams

Prairie in Your Pocket
by Mark Müller

Salamanders in Your Pocket: A Guide to Caudates of the Upper Midwest
by Terry VanDeWalle

The Scientific Nomenclature of Birds in the Upper Midwest
by James Sandrock and Jean C. Prior

Shrubs and Vines of Iowa
by Peter J. van der Linden and Donald R. Farrar

Snakes and Lizards in Your Pocket: A Guide to Reptiles of the Upper Midwest
by Terry VanDeWalle

The Tallgrass Prairie Center Guide to Prairie Restoration in the Upper Midwest
by Daryl Smith, Dave Williams, Greg Houseal, and Kirk Henderson

The Tallgrass Prairie Center Guide to Seed and Seedling Identification in the Upper Midwest
by Dave Williams

Turtles in Your Pocket: A Guide to Freshwater and Terrestrial Turtles of the Upper Midwest
by Terry VanDeWalle

Wetlands in Your Pocket: A Guide to Common Plants and Animals of Midwestern Wetlands
by Mark Müller

Wildflowers of Iowa Woodlands: Second Edition
by Sylvan Runkel and Alvin Bull

Wildflowers and Other Plants of Iowa Wetlands
by Sylvan Runkel and Dean Roosa

Wildflowers of the Tallgrass Prairie: The Upper Midwest, Second Edition
by Sylvan T. Runkel and Dean M. Roosa

Woodland in Your Pocket: A Guide to Common Woodland Plants
by Mark Müller